Paper Cuts:
Recovering the Paper Landscape

JANET N. ABRAMOVITZ
AND
ASHLEY T. MATTOON

Jane A. Peterson, *Editor*

WORLDWATCH PAPER 149

December 1999

THE WORLDWATCH INSTITUTE is an independent, nonprofit environmental research organization in Washington, DC. Its mission is to foster a sustainable society in which human needs are met in ways that do not threaten the health of the natural environment or future generations. To this end, the Institute conducts interdisciplinary research on emerging global issues, the results of which are published and disseminated to decision-makers and the media.

FINANCIAL SUPPORT for the Institute is provided by the Geraldine R. Dodge Foundation, the Ford Foundation, the William and Flora Hewlett Foundation, W. Alton Jones Foundation, Charles Stewart Mott Foundation, the Curtis and Edith Munson Foundation, David and Lucile Packard Foundation, John D. and Catherine T. MacArthur Foundation, Summit Foundation, Turner Foundation, U.N. Population Fund, Wallace Genetic Foundation, Wallace Global Fund, Weeden Foundation, Wege Foundation, and the Winslow Foundation. The Institute also receives financial support from its Council of Sponsors members—Tom and Cathy Crain, Toshishige Kurosawa, Kazuhiko Nishi, Roger and Vicki Sant, Robert Wallace, and Eckart Wintzen—and from the Friends of Worldwatch.

THE WORLDWATCH PAPERS provide in-depth, quantitative and qualitative analysis of the major issues affecting prospects for a sustainable society. The Papers are written by members of the Worldwatch Institute research staff and reviewed by experts in the field. Published in five languages, they have been used as concise and authoritative references by governments, nongovernmental organizations, and educational institutions worldwide. For a partial list of available Papers, see back pages.

REPRINT AND COPYRIGHT INFORMATION for one-time academic use of this material is available by contacting Customer Service, Copyright Clearance Center, at (978) 750-8400 (phone), or (978) 750-4744 (fax), or writing to CCC, 222 Rosewood Drive, Danvers, MA 01923. Nonacademic users should call the Worldwatch Institute's Communication Department at (202) 452-1992, x517, or fax a request to (202) 296-7365.

© Worldwatch Institute, 1999
Library of Congress Catalog Number 99-76556
ISBN 1-878071-51-3

Printed paper that is 50 percent recycled, 20 percent post-consumer waste, processed chlorine free.

Note to readers: As a result of the analysis in this Paper, Worldwatch Institute has reassessed its own paper use and paper buying habits. In 2000 we will be changing the paper used in Worldwatch Papers to one that has a higher post-consumer recycled content and is still processed chlorine free.

The views expressed are those of the authors and do not necessarily represent those of the Worldwatch Institute; of its directors, officers, or staff; or of its funding organizations.

Table of Contents

Introduction	5
Mapping the Paper Landscape	9
Uncovering the Real Costs of Paper	19
Trimming Consumption	31
Improving the Fiber Supply	37
Producing Cleaner Paper	48
Designing a Sustainable Paper Economy	52
Notes	60

Tables and Figures

Table 1: *World's Top 10 Producers of Paper, Share of World Production, 1997*	11
Table 2: *World's Top 10 Consumers of Paper; Share of World Consumption and Population, 1997*	12
Table 3: *Paper Recovery and Use in Top 10 Paper-Producing Countries, 1997*	39
Figure 1: *World, Industrial, and Developing-Country Paper Consumption, 1961–2010 (projected)*	10
Figure 2: *Per Capita Paper Consumption, Various Countries and Regions, 1997*	13
Figure 3: *World Paper Production, Major Grades, 1961–97*	14
Figure 4: *Share of Paper Production in Industrial and Developing Countries, 1970 and 1997*	16
Figure 5: *Sources and Share of Fiber Supply for Paper in the Mid-1990s*	21
Figure 6: *World Fiber Supply for Paper, 1961–97*	22

ACKNOWLEDGMENTS: We would like to express our appreciation to a number of individuals and organizations for their valuable contributions. Many thanks to the following reviewers for their insightful comments on the draft of this paper: Ricardo Carrere, Maryanne Grieg-Gran, Allen Hershkowitz, Susan Kinsella, Yoichi Kuroda, Bruce Nordman, and Maureen Smith. Thanks to the USDA Forest Products Lab (including Said Abubakr, Jim Han, Peter Ince, John Klugness, Gary Myers, Roger Rowell, Tim Scott, Ken Skog, Marguerite Sykes, and Ted Wegner) for providing information and helpful comments on the draft.

During a research trip to Japan, many organizations and individuals gave graciously of their time and expertise. Special thanks to Yoichi Kuroda, of the Institute for Global Environmental Strategies, and to Junko Edahiro. Thanks to the Paper Recycling Promotion Center; Tagui Ichikawa, Shibusawa Masahiro, and Iwamatsu Hiroki of the Ministry of International Trade and Industry; Hisashi Watanabe and the general managers of the Japan Paper Association; and the Japan Overseas Plantation Center for Pulpwood. Thanks also to Nozaki Noa, Hokkaido University; Masako Nakamura, Paper Recycling Citizen's Action Network; and Mr. Konno, Tokyo Metro Area Resource Recycling Companies.

Thanks to colleagues at Worldwatch: Chris Flavin, Chris Bright, Gary Gardner, Anne Platt McGinn, Sandra Postel, and David Roodman for their reviews, to Elizabeth Doherty for artful layout and design, to Jane Peterson for skillful editing, to Sarah Porter for tireless research assistance, to Lisa Mastny for proofreading, and to Dick Bell, Mary Caron, and Liz Hopper for communications and outreach.

JANET N. ABRAMOVITZ is a Senior Researcher at the Worldwatch Institute. She leads the Institute's Forest and Biodiversity Research Team. She is a co-author of the Institute's yearly reports *State of the World* and *Vital Signs*. Her last Worldwatch paper was "Taking a Stand: Cultivating a New Relationship with the World's Forests." She has written about ecosystem services, the causes and consequences of natural disasters, forest products, freshwater ecosystems, and social equity. She was a co-author of the joint Worldwatch/World Resources Institute atlas, *Watersheds of the World*.

ASHLEY MATTOON is a Staff Researcher at the Worldwatch Institute, where she writes about issues related to the conservation of biological diversity. She is a contributing author to the Institute's annual books, *State of the World* and *Vital Signs*. She also writes for *World Watch* magazine.

Introduction

The paper industry can anticipate an extra-white Christmas in the final weeks of 1999. In Japan, 183 banks plan to print out the financial records of all of their customers in order to safeguard against Year 2000 computer glitches. Piled up, these print-outs would tower three times higher than the country's 3,700-meter Mount Fuji. It is uncertain how many other banks, insurance companies, and other financial institutions will follow suit. And a multitude of newspaper and magazine supplements and special editions intended to mark the millennium are also expected to boost paper use.[1]

While printing every customer report, company record, Internet document, or office e-mail seems particularly wasteful, many societies have grown so accustomed to massive uses of paper that this millennial deluge will be only a drop in the ever-rising tide of paper. Soaring paper use has gone largely unnoticed for decades, reaching levels well over 300 kilograms per person per year in some industrial countries—an amount equal to a stack of standard office paper nearly seven meters high.[2]

The paper appetite of industrial countries has not always been so ravenous. Until the mid-1800s, paper was a scarce and expensive commodity made of rags or straw and devoted primarily to printing documents intended to last a long time. The discovery of methods for converting wood to paper cheaply and in mass quantities brought dramatic changes, as did new access to distant forests. Government policies designed to encourage forest exploitation and industrial expansion have also played a major role in fueling the consumption boom. And new uses for paper, such as packaging, copy paper, food service, towels, and sanitary products, have enlarged the market. The average person in the United Kingdom used 16 times more paper at the end of the 20th century than his or her grandparents did at its start.[3]

In the past 50 years, purchases of all sorts of goods, including paper, have skyrocketed along with rising incomes, expanding industrialization, and the promotion of consumption by government and industry. These broad trends, combined with the falling cost of paper, the relatively new ethic of disposability, and high-speed printing and communication technologies have helped fuel a more than sixfold increase in global paper use since 1950. Since the mid-1970s alone, paper use has more than doubled, and there is no end in sight to this trend. World paper consumption is expected to jump nearly a third in the next 10 years.[4]

Not everyone shares equally in the paper feast. With a population only one tenth the size of China's, Japan annually consumes about as much paper as its neighbor. But Japan's per capita consumption is 249 kilograms, while China's is less than 27 kilograms. In the United States, per capita paper consumption is 335 kilograms per year, while in India it is less than four kilograms. Indeed, the vast majority of the world's people live in countries where the average annual consumption of paper is less than the estimated 30- to 40-kilogram minimum each person needs for literacy and communication.[5]

While the impact of a single envelope, magazine, or box may seem negligible, the process of making it requires many steps that take a heavy toll on the world's land, water, and air. Logging, for instance, plays a leading role in global forest loss and degradation—about 20 percent of all the wood harvested is used to make paper. While pulpwood plantations could lessen paper's impact on forests, current plantation practices in many places are hurting forests and local communities. And the making of this product that seems so clean devours vast amounts of chemicals, water, and energy while producing high levels of pollution. The paper industry ranks among the highest in resource use and pollution generation, all to make a product that is usually discarded after being used once. Paper comprises roughly 40 percent of the municipal solid waste burden in many industrial countries.[6]

The challenge—and the prospect—of the new century is to maintain the important services paper provides while lightening the burden that today's heavy paper diet places on the planet. While there have been some substantial gains in recent decades, most have been outpaced by rapid growth in production and consumption. Further improvements in processing will need to be accompanied by reductions in paper use. Production methods and consumption habits in emerging markets present additional concerns as well as opportunities for the future.

There is clear evidence that the challenge can be met. The success of recycling initiatives, which now enable more than 43 percent of the world's paper to find a new life, suggests enormous potential. The 50 percent drop in energy and water use per ton of production in some of the major producing countries also bodes well. The development of products designed to use far less material indicates that manufacturers can use less paper. And because ways to virtually eliminate pollution are known, paper production need not continue to sicken people, plants, and wildlife. So far, most advances have been spurred by citizen demand and government regulation, but industries are increasingly finding that improvements can enhance profitability and create new business opportunities. Thus, the prospect of greater collaboration toward a sustainable paper economy is well within reach.[7]

The start of a new century is a good time to reassess a bedrock assumption of most governments' economic policies—that ever-growing consumption of material goods is essential to economic growth and human well-being. In recent decades, that conventional wisdom has already been challenged with the decoupling of energy use from economic growth, thus raising some fundamental questions about the need for ever-rising use of paper. For instance, how much has this increase actually contributed to our quality of life? How much paper do we *really* need? And how can those needs best be met? In the United States, where per capita use is twice the average for industrial nations, consumption grew by 20 kilograms per person between 1992 and 1997. Yet most people would probably not associate that 20-kilogram increase with improvement in their quality of life, especially if the associated raw material, pollution, energy, and waste disposal costs were taken into account.[8]

Strategies for more sustainable paper use can be integrated into the practices of business, government, and individual consumers alike. Reducing wasteful paper use, redesigning products and practices, and reusing and recycling more paper can bring substantial environmental benefits and save money too. Expanding the use of available environmentally friendly papers—recycled, nonwood, totally chlorine free—will also help transform the paper landscape. Combined, these measures could reduce paper use in industrial countries by one third, trim the world's paper demand, and slash the amount of wood used in papermaking by more than half, while allowing for growth in developing countries to meet basic needs.[9]

Producers also have a key part in reducing the paper burden. New production methods with a proven ability to reduce energy use, pollution, and costs can be employed more broadly than they are today. Shifting the source of fibers for paper from predominantly virgin wood to more recycled and nonwood sources can ease pressure on forests and landfills. The trees used for paper can come from forests and plantations that are managed far better than they are now, causing much less stress on the environment. Felling ancient trees in the world's dwindling frontier forests to make paper need not continue in the new century. Effective standards for forest management, pollution control, and energy use now being followed in some places could be universally accepted and enforced.

Achieving a new paper economy will also require a number of policy changes. A high priority, one that would yield environmental and economic benefits alike, is eliminating subsidies for forest exploitation and virgin paper production. Strengthening regulations governing pollution and waste would help ensure a healthier environment and prevent the disposal of valuable resources. Farsighted policies can encourage the broader use of recycled and alternative fibers and promote the adoption of cleaner technologies instead of discouraging them, as many do now. Governments must also examine the environmental and social impacts that trade liberalization will have at home and abroad before supporting further expansion.

Denuded landscapes, toxic rivers, foul air, bulging landfills, and belching incinerators eventually touch everyone. But the tools are at hand to design a more sustainable paper diet. The paper we do use can be produced in a less harmful way, and high standards of living can be maintained, without incurring high costs.

Mapping the Paper Landscape

"Just imagine a day without paper," an ad in the *Financial Times* for a Finnish paper company reads. "You inevitably see our products every day." And they're right. For the average reader of the *Financial Times*, a day without paper would be almost as impossible as a day without breathing. But in many industrial countries, people are so accustomed to paper—whether it is supplying them with the daily news, drying their hands, holding their groceries, or filling their garbage cans—that its role in their daily lives goes virtually unnoticed.[10]

For most of the 2,000 years since paper was first invented, it was a scarce and valued material, used primarily for important letters and documents. In the last century, however, new technologies, falling costs, and growing economies have allowed the use of paper to skyrocket. Today there are more than 450 different grades of paper destined for purposes as mundane as wiping noses and as specialized as filtering chemicals.[11]

As the uses of paper have expanded, so too has consumption. In 1997, the world's appetite for paper rose to 299 million tons, an amount well over six times the 1950 level. (See Figure 1.) Two hundred ninety-nine million tons of office paper could fill the Empire State Building 383 times or make a pile that could reach the moon and back more than eight times. By 2010, global demand for paper is expected to rise by nearly 32 percent.[12]

Paper use is closely correlated with income levels, and most of the world's paper is produced and consumed in industrial countries. (See Tables 1 and 2.) With 22 percent of the world's population, these nations account for more than 71 percent of paper use. As the populations and economies of

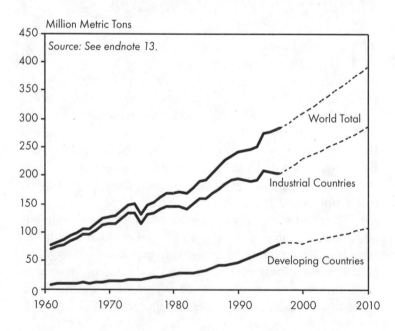

FIGURE 1

World, Industrial, and Developing-Country Paper Consumption, 1961–2010 (projected)

Source: See endnote 13.

many developing nations grow, however, their share of the world's consumption climbs.[13]

On a per capita basis, differences in paper use are even more striking. (See Figure 2.) In 1997, annual per capita paper consumption in the United States was 335 kilograms, and for industrialized countries as a whole, the average was 164 kilograms. But these high-end consumption levels are no indication of the paper diet of the average world citizen. The global average in 1997 was 51 kilograms per person per year; for developing nations it was 18 kilograms. For all of Africa, the average per capita consumption level was less than six kilograms per person, and in over 20 African nations it was below one kilogram. (One kilogram is roughly equivalent to 225 sheets of office paper or two copies of a daily *New York Times*.)[14]

TABLE 1

World's Top 10 Producers of Paper; Share of World Production, 1997

Top 10 Producers	Production (1,000 metric tons)	Share (percent)
World	299,092	
USA	86,477	29
Japan	31,015	10
China	27,440	9
Canada	18,969	6
Germany	15,939	5
Finland	12,149	4
Sweden	9,779	3
France	9,143	3
Korea, Republic of	8,364	3
Italy	7,532	3
Total for top 10 producers	226,807	76

Source: See endnote 13.

TABLE 2

World's Top 10 Consumers of Paper; Share of World Consumption and Population, 1997

Top 10 Consumers	Consumption (1,000 metric tons)	Share of Consumption (percent)	Share of Population (percent)
World	296,896		
United States	89,900	30	5
China	32,695	11	21
Japan	31,374	11	2
Germany	15,733	5	1
United Kingdom	12,240	4	1
France	10,328	3	1
Italy	9,125	3	1
Korea, Republic of	6,836	2	1
Canada	6,652	2	1
Brazil	6,124	2	3
Total for top 10 consumers	221,007	74	37

Source: See endnote 13.

In many parts of the world, expanded access to paper is needed for education, communication, and sanitary purposes. One estimate by the United Nations Environment Program (UNEP) suggests that an annual consumption level of 30 to 40 kilograms per person is essential for education and communication alone.[15]

Paper is used for many different purposes. Today, packaging claims about 48 percent of world paper use. Printing and writing papers make up 30 percent, newsprint another 12 percent, and sanitary and household papers account for 6 percent. While the use of all types of papers has increased, consumption of printing and writing paper in recent years has grown faster than grades such as packaging paper and newsprint. (See Figure 3.) Since 1980, global paper consump-

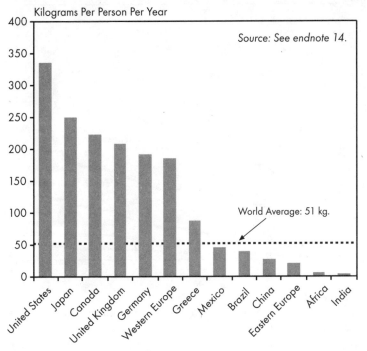

FIGURE 2

Per Capita Paper Consumption, Various Countries and Regions, 1997

tion has jumped by 74 percent while printing and writing paper use has rocketed by 110 percent.[16]

In spite of lighter weight, more efficient packaging, and the replacement of some paper packaging with plastic, the consumption of paper packaging continues to grow. Since the early 1960s, the use of paper for products such as corrugated boxes and food containers has more than tripled.[17]

Decades ago, at the dawn of the electronic information era, many analysts predicted a "paperless" office, but the proliferation of computers, fax machines, and high-speed copiers has instead gone hand in hand with increased use of printing and writing paper. Paper.com, an industry group that examines the correlation between paper use and electronic com-

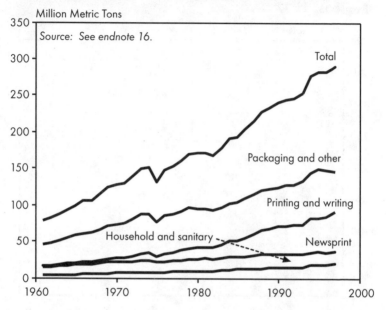

FIGURE 3
World Paper Production, Major Grades, 1961–97

merce, e-mail, and the Internet, describes paper as the "currency of the electronic era," adding, "While the Internet and paper will certainly compete in certain areas, the general pattern is one of mutual growth and interdependence."[18]

Such statements are supported by the trends so far. According to a leading industry analyst, the number of pages consumed in U.S. offices is growing at a rate of about 20 percent each year. In 1996, office copiers in the United States churned out more than 800 billion sheets of paper, laser printers nearly as much. When divided by the size of the U.S. civilian labor force, the amount of office paper consumed was almost 12,000 sheets per person. And electronic mail has not replaced traditional letters either—the number of pieces of mail delivered between 1993 and 1998 increased by 16 percent and the amount of advertising mail rose by 25 percent. The amount of advertising, or "direct," mail in the United States has nearly tripled over the last two decades. In

1998, over 87 billion pieces of "junk mail" were delivered to U.S. households—850 pieces per household.[19]

While some technologies have clearly bolstered the use of certain types of paper, others—such as electronic data interchange—are beginning to cut paper use. Still, it is too early to tell what the long-term impact on world paper markets will be. Information technologies may lead to a decline in paper use eventually, but it is hard to know when, if ever, this will happen. One wild card is the fast-growing markets in some developing countries. As economies grow and technologies spread, the use of printing and writing papers in these countries could skyrocket. But it is also possible that new technologies could allow them to avoid assuming the current wasteful habits of industrial nations.[20]

Rapid changes in the world's pulp and paper industry suggest that the 21st century paper economy will differ greatly from the one we knew in the 20th. Changes in the geography of production and consumption, trade, and investment will have major implications for the landscape and structure of the future pulp and paper economy. For decades, the world's top producers and consumers of pulp and paper have been fairly constant, with the United States, Europe, Japan, and Canada maintaining leading roles. (See Figure 4.) But in the last 10 years, the shares accounted for by these traditional suppliers and consumers have gradually eroded as countries like China, South Korea, and Brazil have emerged as major players.[21]

Nowhere have the changes in markets been more dramatic than in Asia. In the years before the financial crisis hit in 1997, Asia had been home to the world's fastest-growing paper market, increasing at approximately 10 percent a year. Between 1980 and 1997, overall paper consumption in Asia jumped nearly threefold. In Indonesia, consumption rose more than sevenfold, in China more than fivefold, and in South Korea and Thailand well over fourfold. The economic turmoil of 1997 and 1998 resulted in substantial reductions in demand in most of these countries, but by early 1999, markets were showing signs of recovery. Analysts expect

FIGURE 4

Share of Paper Production in Industrial and Developing Countries, 1970 and 1997

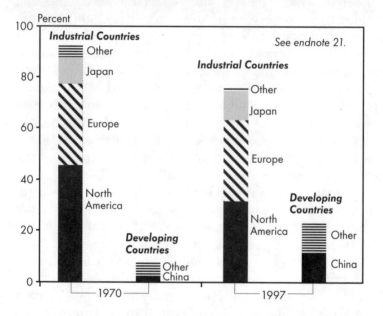

rapid growth to resume again in the near future.[22]

The growth in Asian demand has accompanied growth in the region's production. Between 1991 and 1996, paper production increased by more than 7 percent per year and pulp grew by 8.5 percent. In 1996, Asian pulp and paper production surpassed that of Europe for the first time. And by 2002, some analysts expect that Asia will become the largest producer in the world.[23]

There has also been a flow of investment into Latin America's pulp and paper sector, particularly in Brazil, Chile, Argentina, and Mexico. With the advantages of massive brand new mills, cheaper fiber and labor, and, in some cases, weaker environmental restrictions, production costs for many Asian and Latin American mills are so low that they can easily compete in export markets, even in the United States, Canada, and Europe.[24]

Another change in the final decades of the 20th century was a substantial increase in international trade in pulp and paper. Since the 1960s, the volume of pulp and paper trade has risen more than fivefold. And whereas close to 16 percent of the world's wood pulp and 17 percent of its paper and paperboard were traded internationally in the 1960s, by 1997, these shares rose to 22 and 29 percent respectively. Together these products represent close to 45 percent of the total value of world forest products exports.[25]

For many producers, export markets have driven most growth in the 1990s. Overseas sales by U.S. pulp and paper companies increased by an average 10 percent a year between 1990 and 1997 and accounted for 46 percent of the industry's growth. U.S. companies report that, on average, foreign buyers such as Canada, Japan, Mexico, and Europe account for 24 percent of annual sales. In Brazil, a rapidly growing pulp producer, the leading companies usually sell between 80 and 90 percent of what they make to Western Europe, the United States, and Japan. Pulp exports from Latin American countries are expected to grow by more than 70 percent in the next 10 years.[26]

China is an extreme example of a country that has an enormous and growing influence on global trade in pulp and paper. Economic growth in the world's most populous country has resulted in soaring pulp and paper consumption. Given its population of 1.25 billion, and with per capita use growing at about two kilograms per year, it is no surprise that the eyes of the world's pulp and paper companies are on China. Between 1990 and 1997, paper consumption in China increased by 127 percent while production doubled. In the next decade, the country expects to increase production and consumption by more than half.[27]

To reach these levels, China will need to revamp and expand its industry. Thousands of Chinese mills have been closed in recent years due to outdated technologies and pollution problems, and the country also faces severe raw material shortages. By one estimate, China's wood demand will exceed domestic supply by 40 million cubic meters in 2000, an amount eight times the country's net imports in 1997. While tree-planting efforts and new mills may meet some of the growth in demand, imports of wood, pulp, and paper will also have to increase substantially. Imports have already risen dramatically in the last decade. Since 1990, paper imports have gone up almost fivefold, pulp imports more than threefold. China is now the largest net importer of paper in the world.[28]

One of the more sudden changes in the pulp and paper industry at the close of the 20th century has been the recent trend in consolidation and foreign investment. The unprecedented spate of north-south and east-west mergers and acquisitions in the late 1990s is rapidly changing the paper landscape. The consolidation has been spurred in part by new markets and new producers as well as the need to address the chronic overcapacity and fragmentation problems that have plagued the industry for years.[29]

As mill size and investments have grown over the course of the century, the industry has become inflexible and vulnerable to market shifts. Most of today's new mills are so enormous that a new one can affect the entire global market. A large-scale, state-of-the-art pulp mill costs $1.5–$2 billion to build, and two to three years can pass before it is up and running. By the time a new mill is ready to go, the market may have changed considerably. Yet with the enormous amounts invested, mills must run nearly 24 hours a day, seven days a week to be profitable. A sudden drop in demand is difficult for the industry to respond to, and prices plummet once the market is glutted.[30]

For decades, severe market swings have wrenched the industry, resulting in extreme price fluctuations. Record low prices brought on by the Asian financial crisis in 1997 and 1998 played a role in catalyzing a series of mergers in an effort to improve the industry's dismal financial performance. The recent consolidation trend may make the industry more responsive to market changes and help reduce price volatility by making investment in new capacity more controlled, but it could also make the industry more resistant to change.

Uncovering the Real Costs of Paper

As the world's pulp and paper economy has expanded in size and geographic reach, so too have its associated environmental and health effects. The production of a piece of paper involves numerous steps and subsequent impacts, spanning from soil erosion and species loss when forests are harvested in British Columbia or Chile, to air pollution from pulp mills and waste incinerators in Japan, to the deadly dioxins released by mills along lakes in North America and Russia, to lifeless rivers in China and India. Paper's impacts spread far and wide, and can persist for decades or centuries.

The life cycle of paper usually begins with trees being harvested in a forest or tree farm and transported—sometimes over long distances—to a chip mill, where they are sliced into poker-chip-sized pieces. The chips may then be loaded onto a train or ship and taken to a pulp mill five to 5,000 kilometers away. At the mill, the chips may be mixed with scraps from sawmills, and then dissolved in a slurry of potent chemicals or ground into pulp by mechanical grinders. The pulp is bleached and washed with chemicals and water. It may be shipped a great distance or used in an adjoining paper mill that makes the finished product, such as newsprint or tissue, boxes or writing paper. It will then be shipped to a distributor or printer and eventually purchased and used for shipping other goods, relaying literature or the daily news, or holding coffee for a morning commute.

It is possible that a single piece of writing paper will contain fibers from hundreds of different trees that have collectively traveled thousands of kilometers from their origin in a forest to the consumer's hands. After the paper is used, it has about a 50 percent chance of winding up in a landfill or an incinerator and a 40 percent chance of being recycled—higher or lower depending on the location. Only 10 percent remains in use for a longer period of time, as books or documents filed away.[31]

Perhaps one of the most widely recognized costs of

paper is the threat it poses to the world's forests. Forests are subjected to a barrage of pressures today, and the insatiable appetite for paper is a major one. The world is currently losing about 14 million hectares of natural forest cover each year—an area larger than Greece—and even larger areas are being degraded by less obvious threats such as fragmentation, soil degradation, exotic species invasion, and air pollution. Causes of degradation vary greatly among different regions of the world, but logging for pulp, lumber, and fuel, as well as forest clearance for pasture, farmland, and other forms of development are the leading causes of forest decline.[32]

The virgin wood fiber used to make paper accounts for approximately 19 percent of the world's total wood harvest. Of the wood harvested for "industrial" uses (everything but fuelwood), fully 42 percent goes to paper production. This proportion is expected to grow in the coming years since the world's appetite for paper is expanding twice as fast as that for any other major wood product. By 2050 it is expected that pulp and paper manufacture will account for over half of the world's industrial wood demand.[33]

Even so, the direct connection between papermaking and forest decline is somewhat difficult to sort out. For one thing, the amount of wood used to make paper is often underestimated due to the lack of accounting for sawmill residues. Of the 42 percent of the world's industrial wood harvest going to paper, almost two thirds comes from wood harvested specifically for pulp, while the rest derives from mill residues such as wood scraps and sawdust. In most global statistics, the residues and scraps are not categorized as "pulpwood" and therefore are not accounted for in paper production. In the future, the use of mill residues in papermaking will likely decline and more fiber will come from trees harvested specifically for pulp, as engineered wood products use more of the residues and mill efficiency increases.[34]

Another reason that the link between paper and forests is not always easy to see is that today, just 55 percent of the fibers used to make paper come from virgin wood. Recycled, or "recovered," fibers make up about 38 percent of current total fiber supply for paper, while nonwood fibers such as wheat straw and bamboo contribute about 7 percent. (See Figure 5.) Although nonwood fibers were the primary fiber source for paper until about 150 years ago, their global proportion has remained under 10 percent since the 1960s. The use of recycled fibers, on the other hand, has risen dramatically since that time. (See Figure 6.) Recycled fibers may eventually account for a much larger share of the fiber supply for paper, but for the foreseeable future, the largest share will come from virgin wood.[35]

The sources of wood fiber for paper were fairly constant for most of the time after wood-pulping technologies were invented in the mid-1800s. But in recent decades, wood fiber production has entered a period of rapid flux. The regions, species, and forest types have begun to change.

While the United States, Canada, and Northern European countries have been the mainstays of the world's pulpwood supply in the 20th century, new suppliers in the southern hemisphere have come on the scene and garnered

FIGURE 5

Sources and Share of Fiber Supply for Paper in the Mid-1990s

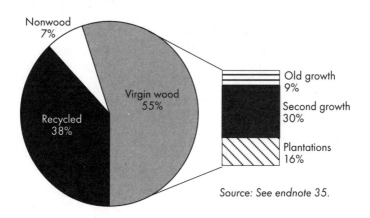

Source: See endnote 35.

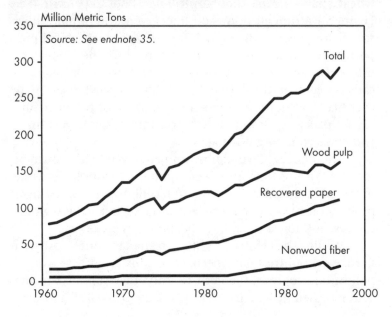

FIGURE 6

World Fiber Supply for Paper, 1961–97

a larger share of the world's pulp production in the last decade or two. Countries such as Brazil, Indonesia, and Chile have realized that they can become global players in the world's pulp and paper market by making the most of strategic advantages—lower production costs and climates conducive to fast growth rates.

In Chile and New Zealand, the widely planted radiata pine grows at about 25 cubic meters per hectare per year, whereas loblolly pine in the southern United States grows at about half that rate. Hardwoods such as eucalyptus—the plantation tree of choice in the tropics—grow at about 40 cubic meters per hectare in Brazil and 26 in Chile, while a comparable hardwood species in Sweden manages a sluggish five cubic meters per year. Faster growth and lower production costs mean that some southern hemisphere producers can provide wood at about half the cost of the traditional suppliers in the North.[36]

In addition to the new geography of pulp, there has also been a significant shift in how trees are grown. Rather than rely on natural mixed-species, mixed-age forests, the industry has shifted toward a more agricultural model where genetic strains are carefully bred and selected, and seedlings are planted in tidy rows and developed into single-species, single-aged stands to be treated with fertilizers, herbicides, and pesticides. These pulpwood plantations are generally harvested in 6–10 year rotations in the tropics and 20–30 year rotations in temperate regions. The uniform, predictable fiber that results is extremely attractive to an industry whose large, expensive mills require a steady flow of easily managed fiber inputs.

In the mid-1990s, pulpwood plantations furnished about 16 percent of the world's total fiber supply for paper. Second-growth forests provided 30 percent, and old-growth forests 9 percent. (See Figure 5.) Of the old-growth forests that are still being logged for pulp, most are in boreal regions of Canada and the Russian Federation. A smaller share comes from original temperate and tropical forests in countries such as Indonesia, Malaysia, and Australia.[37]

While the current harvest from plantations may seem fairly small, large investments in recent decades ensure that the upward trend in plantation sources will continue, and they may eventually become the predominant source of wood fiber for paper. For many countries, the prospect of gaining a larger share of the wood products market has led to heavily subsidized plantation programs and a rush of foreign investment. In Japan, the shortage of domestic timber has led the industry to invest heavily in pulpwood plantations in Chile, Australia, New Zealand, Papua New Guinea, China, Vietnam, and elsewhere.[38]

Today, there are approximately 13 million hectares of fast-growing tree plantations (yielding more than 15 cubic meters of wood per hectare each year), primarily for pulp production. About 80 percent are in South America and the Asia Pacific region. South Africa is also a major player. In Brazil, the industry has been planting around 100 thousand hectares annually with eucalyptus and pine. Similar plantation investments have allowed Chile, like Brazil, to reach the fastest growth rates for industrial wood production in the world.[39]

Proponents argue that intensively managed plantations will create jobs, rehabilitate degraded areas, combat climate change by absorbing carbon, and help "save" forests by providing most of the world's wood needs from a much smaller parcel of land than natural forests might by themselves. A 1999 report from the World Commission on Forests and Sustainable Development suggested that it could be possible to meet the world's demand for pulpwood in 2050 with 100 million hectares of fast-growing plantations—an area equal to more than 70 percent of the amount of the world's cropland planted in corn or three times the area planted in cotton. The implication is that once these 100 million hectares are divvied up, the need to log in natural forests will be reduced and more forests can be protected.[40]

This sounds like an attractive scenario and, to be sure, some types of plantations could play an important role in improving the industry's impact on forests. But plantation development as it is currently unfolding within the pulp and paper industry is not without drawbacks. When compared to degraded farmland, plantations may provide more ecosystem services such as wildlife habitat and soil protection, but when compared to a mature, native forest, they simply don't measure up. Like virtually all large-scale monocultures, plantations are susceptible to disease and pest outbreaks, so they commonly require regular applications of insecticides and fungicides. Herbicides are also used to prevent invasion of competing vegetation. The frequent harvests and site preparation procedures can result in soil degradation that reduces the long-term viability of the land. A mature pulpwood plantation might look like a natural forest, but it actually has about as much in common with a natural forest as a cornfield does with a native prairie.

Another concern is that in some parts of the world, natural forests are being cleared to make way for plantations. In Indonesia, where pulp production has more than quadrupled in the last decade, more than 1.4 million hectares of natural forest have been replaced by plantations. Plantation expansion and the timber industry have been heavily subsidized by the government for years. Satellite data showed that 80 percent of the fires that burned over 2 million hectares of Indonesian forest in 1997 and 1998 were set mainly to clear land for palm oil and pulpwood plantations.[41]

The loss of natural forest in favor of plantations is not only a developing-country phenomenon. In the United States, the expansion of pine plantations for pulp and sawnwood has also come at the expense of natural forests. The United States produces about one third of the world's pulpwood, with most of it grown in the Southeast. According to U.S. Forest Service data, pine plantation cover in the southeastern United States grew by nearly 8 million hectares between 1952 and 1985, while natural pine forest cover declined by 12 million hectares, an area equal in size to the state of Mississippi. The Forest Service predicts that this trend will continue into the future, and anticipates that by 2030 there will be roughly twice as much area in plantation pine as in natural pine stands.[42]

In addition to their environmental impacts, pulpwood plantations have also had adverse effects on local people. In Indonesia, some plantations have displaced indigenous Dayak communities. Companies have failed to negotiate land acquisition agreements with villagers, have broken promises to provide facilities, and at times have harassed and intimidated local people. Similar problems have occurred in many other parts of the world. In Brazil, the Tupinikim and Guarani indigenous peoples have been fighting for decades to have their traditional territories restored. These lands were lost to the Brazilian paper company Aracruz Celulose when it appropriated thousands of hectares of "uninhabited" land in the 1960s.[43]

Another trend threatening forests is the quest for farmland to feed a fast-growing population. In the next 50 years, the world will likely add an additional 3 billion people to its current 6 billion. Already people face severe shortages of arable land in many parts of the world. The United Nations Food and Agriculture Organization (FAO) estimates that an additional 90 million hectares of cropland will be needed by 2010, and at least half of that is expected to come from the conversion of forest. The suggested 100 million hectares that might satisfy human demands for pulp—an increase from the approximately 13 million today—will have to come from somewhere, and there is little productive land to spare. Though generally supportive of plantation development, FAO stated in its 1999 report that "the availability of land for forest plantations is of growing concern.... Large blocks of unencumbered land, even of low fertility, are increasingly difficult to find, particularly in Asia, where the area under forest plantations has expanded most rapidly."[44]

Producing pulp and paper casts a long ecological shadow beyond its impact on the world's forests. To make a ton of virgin paper, at least 2 to 3.5 tons of trees are brought to the mill. Converting those trees into paper uses large amounts of water, energy, and chemicals as well, and generates vast amounts of air and water pollution and solid waste.[45]

Worldwide, pulp and paper is the fifth largest industrial consumer of energy, accounting for 4 percent of all the world's energy use. In Canada, the pulp and paper industry is the largest consumer of energy, in the United States, the second largest. While there have been some improvements in reducing energy use, pulp and paper is still one of the most energy intensive industries in the world—measured as the amount of energy used to make each ton of product—rivaling iron and steel. In developing countries, energy use in pulp and paper is often double that in industrialized countries, because of a greater reliance on outmoded technologies.[46]

In most papermaking nations, energy efficiency has improved in recent decades, due to regulations and cost concerns. In Japan, the amount of energy used per ton of paper has decreased by half since the mid-1970s. In the United States, it declined by 22 percent between 1972 and 1996. However, the United States' efficiency gain was overshadowed by the fact that the total amount of energy used increased by 49 percent, a result of production levels rising by 89 percent.[47]

The pulp and paper industry uses more water to produce a ton of product than any other industry—in the United States, some 44,000 to 83,000 liters per ton of virgin fiber paper, depending on the grade. Writing paper, the fastest-growing grade, uses more water than other grades (three times as much as containerboard, for example) because of extensive bleaching and washing.[48]

Worldwide, pulp and paper is the fifth largest industrial consumer of energy.

Some impressive advances have also been made in water use efficiency. In Japan the amount of water consumed per ton of product has dropped by two thirds in just the last 10 years. In the United States, water use has fallen by 50 to 90 percent per ton since 1950. Some of this progress came as a result of increased use of recycled fiber (which uses less water), some as a result of improved processing methods spurred by regulations.[49]

Converting wood into paper is a complex process that begins by stripping trees of their bark and chipping them into small pieces. Chemicals or mechanical grinders are then used to separate—or "pulp"—the cellulose fibers. The pulp is rinsed and washed several times and often bleached to make it white. Finally, it is formed into paper.[50]

There are a number of ways to accomplish this transformation. Chemical pulping, especially the "kraft" process developed about 100 years ago, is the most common method, used to produce strong papers such as printing and writing paper, wrapping paper and grocery bags, and corrugated boxes. These mills are very large and expensive and can use a wide range of tree species. However, chemical pulping is not very efficient in converting wood into pulp, and only about 50 percent of the wood ends up in paper. The sludgy processing "waste"—the other half of the wood mixed with leftover chemicals—is usually burned to help fuel the mill and recover some of the chemicals. Mechanical pulping takes about twice as much energy as other processes but is about 95 percent efficient in converting the wood into pulp. The paper made from mechanical pulp tends to be weaker and yellows easily, so it is generally used to make newsprint and telephone books, which are not intended to have a long life. There are also pulping processes that combine mechanical and chemical methods. Worldwide, about 40 percent of the pulp that is made is "chemical" pulp, 3 percent is "semi-chemical," 12 percent "mechanical," 38 percent recycled, and 7 percent nonwood.[51]

Pulp and paper mills have long been considered bad neighbors because of the foul-smelling air and sickening water they produce. Even with improvements, in the United States pulp and paper has one of the highest pollution intensities, or emissions per value of output, of the 74 industrial sectors monitored by the government's Toxic Release Inventory. In many countries, mills can be even more polluting, especially if they rely on outdated technology or if pollution is less well regulated. China, India, and other Asian nations, for instance, have thousands of small mills that have no chemical recovery systems, and pour untreated black "pulping liquor" directly into waterways.[52]

Pulp and paper mills also generate solid waste. U.S. mills, for example, produce over 12 million tons of solid processing waste each year which is landfilled, incinerated, and, increasingly, spread on land as a soil additive. While regulatory agencies do not consider this sludge a hazardous waste, it is a source of concern for many local residents and environmental groups because of the residual chemicals (like dioxin) it contains.[53]

A host of air pollutants are released when paper is made, including volatile organic compounds, nitrous oxides, sulfur oxides, acetone, methanol, chlorine compounds, hydrochloric and sulfuric acids, irritating particulate matter, and carbon monoxide. It is the sulfur compounds that give kraft pulp mills a characteristic "rotten egg" smell. In addition to their well-documented human and ecosystem health effects, some of these air pollutants contribute to climate change, and others are ozone-depleting substances.[54]

Thousands of substances are released into water bodies with the waste from pulp and paper mills, including dissolved wood, chemicals, and other compounds—many unidentified—that result from interactions between wood and the pulping and bleaching chemicals. This mix can reduce oxygen levels in the receiving water system and thus kill aquatic organisms, cloud and acidify the water, and spread toxic chemicals. While some life is killed immediately, other effects are long term and persistent as chemicals accumulate and work their way up the food chain to people. Some of the most deadly (and avoidable) compounds released are the chlorinated organic compounds, grouped under the heading "AOX" for adsorbable organic halides. This group includes dioxins, furans, and chloroform.[55]

Many of these highly toxic compounds are produced during bleaching processes that use chlorine. To make whiter paper, wood pulp is bleached, most commonly with elemental chlorine (usually in the form of chlorine gas). Chlorine bleaching has come under intense scrutiny since the discovery of dioxins in mill effluent in the mid-1980s. Some mills have switched to other types of chlorine, such as chlorine dioxide, which can cut measurable AOX discharges by 90 percent. Worldwide, about half of the total bleached production now uses chlorine dioxide, and is called Elemental Chlorine Free (ECF). The use of the ECF label has come under some criticism because it can give the false impression that no chlorine is used, which is not the case. And many of the chlorine-based compounds are toxic at levels that are too low to detect using standard measures.[56]

There are, however, many bleaching chemicals available that do not produce the highly toxic chlorine byproducts, and some mills are using them. These safer alternatives include oxygen, hydrogen peroxide, and ozone, and are labelled Totally Chlorine Free (TCF).

Most of the improvements in environmental performance in industrial countries have come in response to government regulations. The earliest regulations tended to deal with end-of-the-pipe (or smokestack) pollution and to prescribe specific technologies. In the United States, beginning in the 1950s, a series of laws were enacted to deal with various forms of pollution—air, water, solid waste, toxics, and so forth. The Clean Water Act, Clean Air Act, and other laws were successful in reducing the gross levels of pollution that characterized the pulp and paper industry of 30 to 50 years ago, when dead and discolored rivers and lakes were commonplace.

Today, there is greater emphasis on pollution prevention or source reduction. This newer approach tends to set longer-range emissions targets with phase-in periods, and allows industry to figure out how best to meet these goals by examining and modifying its production processes to reduce pollution. As the public, regulators, and the industry have become more sophisticated, there is a growing awareness that the most effective way to achieve environmental and economic goals is to prevent pollution in the first place: wasted resources are wasted profits. In Sweden, strict standards with flexible rules for compliance have helped that country achieve higher environmental performance and greater competitiveness. The recent U.S. Environmental Protection Agency (EPA) "Cluster Rule" also takes a more integrated approach to air and water regulations by allowing industry to meet all regulatory requirements simultaneously, and gives them flexibility in selecting the methods and technologies they will use. The EPA estimates that the rule will cut the amounts of various air pollutants to roughly half of current levels, chloroform levels by 99 percent, and dioxins and furans by 96 percent.[57]

Environmental improvements have not been uniform around the world. In most developing countries, existing paper mills are highly polluting as there is little or no pollution control technology. Of particular concern is the rapid growth of the pulp and paper industry in developing countries that have weak environmental standards and enforcement capability. The few standards that exist tend to focus on concentration of pollutants rather than the amount of pollutants—encouraging dilution as the way to tackle pollution.

As a result, pollution is not cut, and water use is excessive. Further complicating efforts to clean up pollution in developing countries is the fact that the agencies responsible for the task tend to be grossly underfunded and understaffed.[58]

Trimming Consumption

While the industry itself will necessarily play a key part in reducing the environmental impacts of paper and shifting toward sustainability, consumers also have a pivotal role. It is their purchases and preferences that send signals to the industry. And consumers' discards help create the mountains of waste, much of which is made up of paper. Strategies such as "reduce, redesign, reuse, and recycle" can easily be practiced by business, government, and individual consumers alike.

Business consumers can exercise particularly strong influence over the industry. Office paper is the fastest-growing use of paper, and the cost of printing, copying, mailing, storage, and disposal can exceed the initial purchase price by as many as 10 times. There are easy ways for businesses to audit and reduce their paper use and costs. Reductions of 20 percent or more are possible in most offices, and have been achieved in many, through "good housekeeping" practices such as eliminating "extra" paper purchases and needless copies and reusing paper that is blank on one side. Photocopying and printing on both sides can save a substantial amount of paper, and office machines that print "duplex," or double sided, are widely available. When combined with printing two pages per side (especially good for archiving), the amount of paper used in photocopiers can be reduced by up to 75 percent. Duplexing can also reduce the weight and the cost of sending documents through the mail.[59]

Lighter-weight and smaller-sized papers provide another option. In the United States, the standard office copy paper is 20-pound paper (a label that refers to the weight of

2,000 sheets). The Lawrence Berkeley Lab has calculated that changing 75 percent of the 4 million tons of copy paper used in the United States each year to 18-pound weight would trim paper use by 300,000 tons per year. In Europe, the typical paper is slightly larger and heavier (about 6 percent heavier), so there the savings would be even greater. (Japanese paper, on the other hand, tends to be slightly lighter than U.S. paper).[60]

Today's new generation of information and communication technologies is showing how businesses can function with far less paper. Electronic mail, electronic data interchange, document scanners, intranets, and the Internet can radically reduce paper use, while also saving time and money. Many companies are now producing and processing core business documents like invoices, purchase orders, and reports electronically, greatly increasing their efficiency.

Many companies have already begun using paper-saving strategies. Bank of America, now the largest commercial bank in the United States, decided in 1994 to reduce its paper use by an ambitious 25 percent in just two years. It reported that it exceeded that goal by 1997 by using online reports and forms, e-mail, and voice mail instead of paper forms and memos, duplex copying, and lighter-weight papers. Trimming the weight of the paper in automatic teller machines by 25 percent alone saved 228 tons of paper a year. Bank of America now recycles 61 percent of its paper, saving about half a million dollars a year in waste hauling fees. And 75 percent of all its paper purchases have recycled content.[61]

Many printers and publishers have managed to reduce the amount of paper waste generated at the printing plant and recycle the waste they create, saving money in the process. Advances in computer-assisted layouts and other design tools allow printers to maximize the amount of printing on a sheet. Some printers are using lower-weight papers. The weight of the paper in American newspapers, for example, was gradually reduced by 9 percent from the late 1960s to the mid-1980s. Some newspapers and magazines are saving even more money and paper by also reducing the size of their publications by a fraction on each side.⁶²

It is still common for magazine publishers, however, to print and ship many more copies than they expect to send to subscribers or sell at the newsstand. At least half of the magazines printed in the United States are never sold. Although most of the overruns are recycled, reducing these overruns can bring significant resource savings.⁶³

The rapidly growing overnight and express shipping industry is another large consumer of paper. United Parcel Service (UPS), the largest such company in the world, ships over 3 billion packages a year. Together, the five largest companies have yearly revenues of nearly $100 billion. There are ample opportunities for shippers to reduce the amount of material they use and save money by using lighter packaging and redesigning their shipping materials for easy reuse. They can also expand their recycling efforts and raise the recycled content and recyclability of their materials. Some companies (such as Airborne, UPS, Federal Express [FedEx], and the U.S. Postal Service) now use more than 80 percent post-consumer wastepaper for boxes and paperboard, are eliminating bleached paper, and offer reusable materials.⁶⁴

> **Today's new technologies can radically reduce paper use, while also saving time and money.**

Smaller businesses can make a difference as well. Thanks to an initiative started in 1990 by a concerned office worker, an increasing number of Japanese offices are coming together to collect paper in sufficient quantities to make its pickup by a private recycling company economical. One such "office town council," a group of 280 offices, now diverts more than 8,000 tons of high-grade office paper from the waste stream each year, saving nearly 87 million yen (approximately $712,000) in disposal costs. The group has also promoted purchases of recycled content paper products. Their example has been replicated elsewhere in Japan.⁶⁵

Each household may not use as much paper as a business, but collectively their paper use is considerable, as is the influence they have on businesses that provide them with goods and services. Guides are now available to help consumers audit and reduce their own paper use, and identify and support more benign products. They can eliminate excessive "junk mail" and catalogues, choose items with less packaging, and support community recycling efforts. An essential step is buying recycled products, for without a market, paper that is collected for recycling may be end up in a landfill or incinerator.[66]

Some manufacturers are finding creative ways to produce the same desired product while using less raw material. In the 1980s, for example, standards for corrugated shipping containers shifted from weight-based standards to performance-based standards. By "lightweighting," a substantial amount of raw material was saved, yet stronger containers resulted. Since 90 percent of materials shipped in the United States use corrugated packaging and about half of the world's paper is turned into packaging and shipping products, this shift has been significant.[67]

Products and services can be redesigned to use far less material or material less harmful to the environment. Consumer products giant Procter and Gamble shaved the amount of paper packaging per unit of product by 24 percent in just four years. In Europe, some padded shipping envelopes are a "bag-in-a bag" that can easily be reused and recycled. The plastic bubble-wrap bag slips out of the paper envelope and both can be recycled.[68]

Entire products and functions are being redesigned to virtually eliminate paper use. Documents such as phone directories, parts catalogues, technical reference manuals, and company reports can now be accessed online or on CD-ROMs, saving millions of dollars and tons of paper. The potential for saving paper, money, and time is illustrated by the fact that the contents of all the phone directories in the United States could be put onto a few CD-ROM disks for just a few cents, and are already available on the Internet.[69]

While the paperless office predicted by some in the

1970s has not materialized, today's innovations are beginning to trim paper use. A Danish company, for example, cut its paper purchases by about half, just by expanding its use of e-mail. Other technologies—such as electronic books or the "decopying machine" that removes toner from old paper so that each sheet can be used several times—show that redesign is possible. While not all of these innovations may be widely adopted, they illustrate possibilities for revolutionizing communications and paper use in the coming decades.[70]

Paper products from office copy paper to corrugated containers and shipping envelopes to magazines can be reused before recycling. UPS is introducing reusable shipping envelopes that not only cut down on materials but also make it easier for the recipient to respond to the sender. Manufacturers of a range of products like furniture and computers are also switching to returnable and reusable packaging, which can save them money.[71]

Municipalities and businesses have found that expanding recycling has reduced the volume of waste. Retailers that receive large shipments have already found that recycling old corrugated containers can be profitable, and their efforts help account for the fact that corrugated paper has the highest recovery rates of all paper types. In the United States, over 70 percent of old corrugated containers have been recycled since 1995, and grocery stores are now the single largest source of this valuable material.[72]

As very large consumers of paper, publishers can demand that the market supply better papers. Dutch paper buyers, for example, are demanding old-growth-free paper from Finland, and German publishers are making sure that their suppliers practice good forest management. Printers are finding that customers are asking for these papers, and firms that supply them have a competitive edge. Members of like industries, such as publishers or environmental organizations, have come together to form buyers' groups so that collectively their purchases can help spur change.[73]

Governments are also big paper purchasers. The U.S. government, which accounts for 2 percent of all paper bought in the country, has mandated that its purchases must have a minimum of 30 percent post-consumer recycled-content. Japanese government offices now buy recycled content copy paper with whiteness of 70 percent (instead of the old standard of 80 percent), thanks to a pro-recycling campaign by concerned businesses and nongovernmental organizations (NGOs). The campaign also taught consumers, including the government, that whiteness standards for paper were unnecessarily high.[74]

A number of creative alliances are bringing changes. The U.S. EPA set up the WasteWise program five years ago to encourage public and private entities to reduce waste by sharing strategies and highlighting achievements. Last year, more than 900 partners cut waste by 8 million tons and saved about $40 million in avoided purchases and disposal fees. The overnight shipping industry and the fast-food giant McDonald's have worked with the Environmental Defense Fund to reduce their materials use. The U.S. Postal Service is collaborating with the U.S. Department of Agriculture's (USDA) Forest Products Lab (FPL) to develop better stamp adhesives that will make paper more recyclable. By using the FPL's innovative capacity and their own position as a large paper buyer, the postal service is hoping to help shift the paper industry to more sustainable practices.[75]

In recent years, many advances have been made in the quality and availability of more "environmentally friendly" paper. The perception of recycled paper, for example, as weak or coarse and flecked is no longer true, as a range of high-quality recycled paper that meets the demands of all types of printing is available at competitive prices. Likewise, the range and availability of "tree-free" papers and those made without chlorine bleaching have expanded considerably, and guides are available to help consumers make those choices and find suppliers. Labels that indicate recycled content, chlorine-free bleaching, and so forth are becoming widespread in many parts of the world. Consumers are using this information to buy better products and encourage the industry in a sustainable direction.[76]

Improving the Fiber Supply

While significant progress has been made in reducing the environmental costs of producing paper in many parts of the world, there is still room for improvement, and considerable responsibility for easing the planet's paper burden falls to producers. An important step is to improve the mix of fibers in paper. Virgin wood fiber need not continue to be the mainstay of paper manufacturing. Expanding the use of alternatives, such as recycled and nonwood fibers, and improving forest management practices could greatly reduce the real costs of producing and processing the basic material for paper.

Expanding the collection and reuse of old paper is one of the most promising ways to lessen many of the problems associated with paper—by reducing the pressure to cut more trees, easing overburdened waste disposal systems, and cutting energy use and pollution, to name a few. As its benefits have been recognized, recovered paper's contribution to the global fiber supply for paper has nearly doubled, from 20 percent in 1961 to 38 percent 1997.[77]

Producing new paper from old is an efficient process. For each ton of used paper, nearly a ton of new can be produced—far more efficient than the 2 to 3.5 tons of trees used to make one ton of virgin paper. And because recycled paper has already been processed, far less energy and chemicals are required during reprocessing, just 10 to 40 percent of the energy consumed for virgin pulping. Much less bleaching is needed because white papers have already been bleached during their original production. Many grades (like corrugated boxes) do not even need bleaching.[78]

Recycling can make use of a huge, barely tapped supply of materials—the "urban forest." This term can be used to describe cities, not because they grow trees, but because they generate enormous amounts of wood and paper waste that is all too often thrown away. A fast-growing industry is making use of this vast resource to produce useful products, reduce waste, and create jobs. A mill in the Washington-Baltimore corridor takes 700 tons of mixed office waste paper each day and converts it into a high-quality pulp that can be used for printing and writing paper. The clean process uses no chlorine and recycles its water, producing far less air and water pollution and using far less energy than the virgin pulp mills its product competes with. And this recycled pulp is priced competitively. A new paper mill is under construction at an old industrial site in the Bronx in New York City that will use 100 percent recycled newspaper and get its water from treated sewage. It will produce far less pollution (99 percent less nitrous oxides, sulfur oxides, and volatile organic compounds) and use half the energy of virgin production. Locating near the source of its raw material and the buyers for its products will slash transportation energy costs by 94 percent, and it is expected that the paper will cost 28 percent less than virgin newsprint.[79]

In many countries, the primary motivation for increasing recovery rates has been the need to reduce the flow of waste to landfills and incinerators. The volume of waste generated in many industrial countries has grown substantially in recent decades, more than doubling in the United States alone since 1960. Paper accounts for the largest share of municipal solid waste (MSW) in many industrial countries. In the United States, for example, paper makes up 39 percent (by weight) of MSW generated. Even though almost half is now diverted for recycling, some 44 million tons are discarded each year—more than all the paper consumed in China.[80]

As the benefits of recycling have been recognized, and governments have promoted it, paper recovery has grown. Between 1975 and 1997, the volume of paper recovered worldwide more than tripled, from 35 million to nearly 110 million tons. During that time, the wastepaper recovery rate—the share of paper used that is recycled or recovered—increased from approximately 38 percent to more than 43 percent. FAO predicts that by 2010, global consumption of recovered paper will reach 177 million tons, with a projected recovery rate of 45 percent.[81]

Wastepaper recovery rates vary dramatically among countries. (See Table 3.) Legislation to aggressively reduce solid waste in Germany has resulted in paper recovery rates of nearly 72 percent. And in Japan, the world's second largest paper producer, limited domestic resources and a shortage of waste disposal options have encouraged the heavy use of recovered paper. Indeed, the Japanese paper industry itself helped establish the Paper Recycling Promotion Center and has set a target of 56 percent.[82]

Both mandatory laws and voluntary targets have been very successful in expanding recovery and recycling. In the

TABLE 3
Paper Recovery and Use in Top 10 Paper-Producing Countries, 1997

Country	Recovered Paper/Paperboard			Recovery Rate[1]	Utilization Rate[2]
	Total Recovered	Exports	Imports		
	(1,000 metric tons)			(percent)	(percent)
World	128,725	16,460		43	44
United States	40,909	6,823	630	46	40
Japan	16,546	312	362	53	54
China	8,760	4	1,618	27	38
Canada	3,110	688	2,088	47	24
Germany	11,279	2,739	918	72	59
Finland	607	49	84	35	5
Sweden	1,323	193	559	55	17
France	4,270	750	998	41	49
Korea, Republic of	4,530	0	1,452	66	72
Italy	2,784	53	926	31	49

[1]Total recovered paper volume divided by apparent paper and paperboard consumption. [2]Total recovered paper consumption divided by paper and paperboard production.
Source: See endnote 84.

1970s and early 1980s, only about one quarter of wastepaper was recovered in the United States. Thanks to a variety of laws and private initiatives (such as banning paper in landfills, establishing curbside recycling programs, and issuing mandates for recycled content paper), recovery rates rose to 46 percent by 1997. In Europe, which also has high levels of wastepaper recovery, a 1994 European Union Directive targeted a recovery rate of 50 to 65 percent for packaging waste by 2001, and a new directive calls for a nearly two-thirds reduction in the amount of biodegradable material (such as paper) sent to landfills. Combined with expanded recycling programs, these laws will reduce waste and increase paper recovery. The Netherlands is on the way to meeting its goal of recovering more than 72 percent of the paper sold inside its borders by 2001.[83]

Because used paper is traded between nations, recovery rates do not necessarily indicate the amount of old paper a country actually uses to produce new paper. Fifteen percent of all recovered paper entered world trade in 1997. Although Sweden recovers well over half of what it consumes, the country is such a large producer and exporter of paper that the relative contribution of recovered paper to overall paper production is only 17 percent. In the United States, the largest exporter of used paper, rates for utilization of wastepaper remained close to 23 percent between 1965 and 1985. By 1997, they reached 40 percent, a level not seen since the 1940s. (See Table 3.)[84]

While recycling has slowed growth in the demand for wood pulp, it has served more as a supplement than as a substitute for virgin fiber. Global paper and paperboard consumption has been increasing so rapidly that it has overwhelmed gains made by recycling. So while the amount of material recovered has increased sevenfold since 1961 and its share of the fiber supply has nearly doubled, the total volume of virgin wood pulp and paper consumed and waste generated continues to rise, overtaking these important successes.[85]

The potential for using old paper to provide a steady stream of fiber for new paper has yet to be fully exploited.

Today's 43 percent recovery rate is far below the 70 percent or more of old paper that could be recycled. Some grades of waste paper, such as old corrugated and newspapers, are more widely recycled than others, and there are well-developed markets for pulping them to make new like products (such as new corrugated boxes made from old corrugated boxes). Other grades, such as mixed office paper, have lower recovery rates and very little of what is collected is used in making new office paper. Instead, it is downgraded for other uses such as cardboard because of the variety of inks used and the demand for ultra-bright white office paper. In fact, more than 90 percent of the printing and writing paper made in the United States is from virgin fiber, only 6 to 7 percent from recycled. In the United Kingdom, which has scarce raw materials and imports two thirds of its newsprint, only 40 percent of its old newspapers are recycled, while the rest are landfilled. Their Newspaper Publishers Association recently issued a report confirming that expanded recycling would provide considerable environmental benefits and improve the industry's competitiveness.[86]

Paper consumption has been increasing so rapidly that it has overwhelmed gains made by recycling.

A number of hurdles—some technical or economic, some social—prevent recovery and recycling from reaching its full potential. One barrier has been price volatility in recovered paper. Until recently, the supply of recovered paper fluctuated widely, as did prices, as collection programs were adopted and abandoned, then re-adopted in response to prices. Volatile global markets also affected prices. For many years, the capacity to pulp wastepaper was limited, and the volatility of supply and price made it difficult for mills to invest in new facilities to handle the paper, creating a vicious circle. Now, with the more widespread adoption of municipal and business recycling programs and significant expansion of mill capacity for recycled material, supply and prices are more predictable. Still widespread subsidies for virgin fiber production and for landfills and incinerators continue to put recovery and recycling at an economic disadvantage.[87]

Despite the expansion and success of recycling programs, their economic benefits are usually underestimated. As a result, municipalities tend to underinvest in recycling. In Massachusetts, these economic benefits are two to four times the waste disposal fees. The additional investment needed to fully expand recycling would amount to just 1 to 2 percent of the economic benefits that would accrue to municipalities.[88]

Many of the technical barriers that once stood in the way of using more recycled paper—such as insuring fiber strength, removing new types of printer and copier inks, and eliminating "stickies" (adhesive labels and such)—have been solved or are close to being solved. In addition, some new technologies are increasing the strength and recyclability of recovered fibers. Still more work is needed on improving the recyclability of mixed office paper, the fastest-growing segment of the paper market.

What may be the biggest hurdles to expanding recycling in some industrial countries are the persistent myths about recycled paper. Some articles have mistakenly claimed, for example, that recycling paper uses far more energy than making virgin paper, or that the quality of recycled paper is too low to meet printing standards and consumer preference. The truth is that even though collecting old paper may use slightly more energy per vehicle than collecting trash, when the total energy budget of recycling is compared to virgin production, recycling wins by a landslide because virgin processing is so energy intensive. And of course recycling also saves forests and water.[89]

Many skeptics also claim that recycled paper cannot meet the same standards as virgin papers. But in recent years there have been dramatic advances in the quality of recycled papers, thanks to innovations in processing (such as enzymatic deinking to remove stubborn toner inks). And as the volume of recovered paper use rises each year, even more gains are being made. The most common standards for judging writing papers—opacity and brightness—are easily met by today's recycled papers. The strength of recycled paper as well is on a par with virgin paper (a concern for printers because breaks in large paper rolls can be very costly). And in many surveys, consumers have a stated strong preference for recycled.[90]

Nonwood fibers could also contribute to a more environmentally benign fiber mix. There are three main types of nonwood fibers for paper: agricultural byproducts from crops such as wheat, rice, and sugar; crops such as kenaf and industrial hemp that can be grown specifically for pulp; and wild plants like reeds and grasses. Textile scraps are also used as a fiber source in some places, but in much smaller amounts.[91]

While nonwoods currently make up about 7 percent of the world's total fiber supply for paper, there is a strong case for increasing the share of nonwood fibers to 20 percent or more. The incorporation of a larger share of nonwood sources could make use of a resource that is currently burned in many parts of the world, provide farmers with an additional source of income, reduce chemical use in pulping, and drive down the demand for wood pulp.

Nonwood fibers were once the sole source of raw material for pulp; it wasn't until the middle of the 19th century that wood-pulping techniques were even invented. The first piece of paper, produced in China in 105 A.D., was made of tree bark, old rags, hemp, and used fishing nets. As recently as the 18th century, used rags and cloth were the primary source, but as demand began to outstrip supply in the early 1800s, a search for new sources began. Since the dawn of the wood-pulping era, the use of nonwood fibers in papermaking has gradually declined to its current marginal share.[92]

Developing countries account for 97 percent of the world's nonwood pulp use. China consumes 83 percent of global nonwood pulp, and India accounts for nearly 5 percent. In the United States, nonwood fibers account for less than 1 percent of the paper industry feedstock, whereas in China nonwood fibers (primarily straw) make up nearly 60 percent. In the future, China may rely more on wood and recycled paper and less on nonwoods, largely because nonwoods have been processed in older, smaller mills that are inefficient and cannot afford important pollution-abatement measures. In recent years, thousands of these mills have been closed, and large wood-based mills are likely to replace them.[93]

There are many reasons to consider expanding the role of nonwoods in the papermaking process, not to replace wood sources entirely, but to displace a portion of the wood used for paper. One of the primary advantages that nonwoods have over wood is their low lignin content. Lignin binds cellulose together and the removal of lignin is part of what makes the pulping process so energy and chemical intensive. While lignin makes up 23 to 34 percent of the total chemical composition of wood fibers, nonwood sources such as rice straw and kenaf have a lignin content of 9 to 15 percent. The lignin content of industrial hemp can be as low as 3 percent. For many nonwood fibers, the lower lignin content corresponds with a higher proportion of cellulose. While wood fibers are less than 50 percent cellulose, cotton linters (byproducts of cotton processing) are 85 to 90 percent cellulose—making them an extremely valuable papermaking material.[94]

A major drawback for some nonwood fibers is their high percentage of silica. This is particularly problematic for agricultural residues. While the amount of silica in wood fibers is negligible, silica content for different straws ranges between 0.5 and 14 percent. The silica is difficult to remove in pulp mills' conventional chemical recovery process. It can therefore result in a highly polluting effluent and costly losses of chemicals and water that could otherwise be recycled. However, there are ways to get around the silica problem. Since up to 50 percent of the silica in many straw fibers comes from soil residues, washing the fibers before pulping can reduce the silica load dramatically. In addition, there are alternative pulping techniques that enable the recovery of chemicals.[95]

Agricultural byproducts such as wheat straw, sugar cane bagasse, and corn stalks currently make up about three quarters of the world's nonwood pulp supply. By one estimate, one hectare of cereal grain can yield up to one ton of straw—and this includes an allowance for half of the straw to be plowed back into the soil. The total availability of agricultural byproducts could be over 2 billion metric tons per year, far more than the roughly 35 million tons currently used.[96]

The majority of agricultural residues should be composted and recycled on farmlands, and in some countries, residues represent an important source of fuel. Yet in some parts of the world, large amounts of these residues are burned, resulting in polluted air and a wasted resource. By incorporating a small share of this material into the fiber supply, the industry could provide an additional source of income for farmers and make use of a material that may already be conveniently collected at processing mills.[97]

Crops like kenaf and hemp that are planted specifically for pulp also have potential for expanded use. These sources currently make up about 9 percent of the world's nonwood pulping capacity. They can cost more than agricultural byproducts, but many have properties that yield high-quality pulps as well as environmental and social benefits. In some places, kenaf and hemp crops can produce greater yields than typical tree plantations, and the lower lignin content of their fibers makes the pulping process less chemical and energy intensive.[98]

Some critics argue that planting crops of kenaf or hemp could be more environmentally harmful than planting crops of trees. Assuming similar management practices, both would be monocultures, could contribute to soil erosion, and would be vulnerable to pest outbreaks and disease. And as far as wildlife habitat and watershed protection values go, a 20-year-cycle pulp plantation would likely come out ahead of an annual crop. But beyond lower chemical and energy requirements and a potential for higher yields, industrial nonwood crops offer additional social and economic benefits that pulp plantations do not.

Ultimately, whether kenaf or hemp is preferable to wood fiber is highly dependent on the local social, economic, and ecological conditions of a region. In some places, the climate may be more suitable for nonwood crops; in others, farmers may be looking to diversify crop options and could get a much faster return from an annual crop than from a pulpwood plantation that could take 15 to 20 years to mature.

Some of the main obstacles to expanding the use of nonwoods are industrial inertia and over 100 years of investment in forest resources and billion-dollar wood-pulping mills. In addition, current economies of scale and subsidies for timber harvest and production make it difficult for small-scale, localized nonwood mills to compete.

Even with expanded use of recycled and nonwood fibers, a substantial share (25–30 percent) of the world's fiber supply for paper will likely come from virgin wood. Tree plantations and forests are integral to the modern paper production system, but in a world with a rapidly expanding population and a declining forest endowment, reforming forest management is essential. Continuing the practices of the 20th century combined with a growing demand for wood would leave us with severely degraded, fragmented systems 50 years hence. Large expanses of healthy, intact forest ecosystems that are not in protected or extremely remote areas would likely become a distant memory, and vital ecosystem services would be severely compromised.

There are some encouraging trends that demonstrate how forest management can be improved. In recent years, there has been a growing interest in sustainable forestry practices on the part of local communities, foresters, industry, policymakers, and concerned citizens. These practices involve changing harvesting techniques, managing for multiple species, age classes, and uses, and protecting wildlife habitat and watersheds while also providing products and livelihoods.

Some types of plantations can play a role in reducing the environmental impacts associated with the production of pulp for paper. Farming trees in a sustainable way is clearly preferable to harvesting the world's last remaining old-growth stands. But in general, plantations can be managed much better than they are now. It is important that they be established on lands that truly are degraded—that are not currently forested, farmed, or inhabited, and do not have high potential to regenerate naturally. Instead of providing subsidies for plantation establishment in recently cleared or highly productive areas, governments could offer financial incentives to plant in degraded areas. These subsidies could make up for some of the profits lost from slower growth rates.

Other ways to improve plantation management include protecting watersheds, involving local stakeholders, reducing fertilizer and pesticide use, using native trees, and planning land use for the long term. Some companies have already taken steps to reduce their environmental impact. For example, a Brazilian company intersperses its plantations with plantings of native rainforest species. They do this in part to improve their public image, but also to provide a natural means of pest control.[99]

In some parts of the world, consumer demand for sustainably harvested products has played a role in encouraging the trend toward sustainable forest management. Increased consumer awareness, especially in the European market, has recently caught the attention of major forest products industries. And companies ranging from Dutch publishers to McDonald's have recently begun demanding products free of old-growth wood.[100]

The growing concern over forest management practices has contributed to an expansion of certification initiatives in the 1990s. The Forest Stewardship Council (FSC) sponsors the best known and most credible certification program. FSC accredits certifiers who, at the request of companies wishing to use the FSC logo, audit forest management practices and certify products for the entire chain of custody, from forest to transport to processing. FSC-certified forests must follow strict standards set forth in regionally specific principles and criteria for sustainable forest management. Many companies are seeking this certification, and by the beginning of 1999 over 15 million hectares of forest—equal to an area nearly three times the size of Costa Rica—had been certified by FSC-accredited certifiers. In late 1998, the first U.S.-produced paper containing FSC third-party-certified wood pulp arrived on the market.[101]

In addition to the FSC, there are many other certification initiatives under way. However, they lack similarly stringent criteria and indicators for forest management and reliable third-party certification processes. As certification programs gain steam, it will be important for consumers to learn about the real meanings of the different labels they encounter so they can encourage the most reliable initiatives.[102]

Producing Cleaner Paper

Regardless of the fiber source, there is ample room to improve the way that paper is made. In recent years, new methods have been developed for producing paper with less water, less energy, less pollution, and less virgin raw material. Some businesses have begun to embrace these changes. Many companies were initially prodded by regulations, but a growing number are finding it more profitable (and publicly acceptable) to generate less waste and produce more environmentally benign products. On the whole, however, the pulp and paper industry has been rather slow to change. And relatively few companies have been willing to adopt innovations that fail to pay for themselves in the next quarter.

Reusing water, chemicals, and other materials in a manufacturing operation is a time-tested way of increasing profitability and improving environmental performance. A logical extension of this approach is the development of "minimum-impact mills" that keep natural resource consumption (wood, energy, water) as well as releases to the water, air, and land as low as possible. The ultimate goal is the complete elimination of these releases through a "closed-loop" or "zero-discharge" system.[103]

There are already several zero-effluent mills in the world. One is a mini-mill located in the New Mexico desert that uses old corrugated containers to make 100 percent recycled linerboard (for such items as boxes). While similar plants consume 2.5 million gallons of water a day, this mill requires only one fifteenth that amount, taking steam and water purchased from a nearby electric plant and recycling it over and over in a closed loop. A Wisconsin recycled paper mill has completely eliminated discharge. This mill converts 98 percent of the fiber into new containerboard, ranking it among the most efficient in the industry.[104]

As noted, progress has been made in reducing the volume of waste produced in pulp and paper mills, and the levels of some of the most toxic chemicals like dioxins. Using better methods for removing lignin (such as extended cooking and oxygen delignification) that reduce chemical and energy use and give higher fiber yields, switching to safer bleaching chemicals, and better waste treatment are major steps toward cleaner paper. Modern mills that have adopted improved processes have cut effluents by 80 percent, which in turn helps cut energy and chemical demands, saves money, and makes it easier to comply with environmental permit requirements.[105]

A number of other new technologies have the potential to significantly improve pulp and paper processing. "Biopulping," for example, takes advantage of naturally occurring local fungi to break down lignin in the wood chips, thereby reducing the amount of energy needed in processing and improving the water quality of effluents, while increasing the strength of the resulting paper. Researchers at the USDA Forest Products Lab are developing "fiber loading" as a way to produce lighter-weight printing and writing papers from recycled or virgin pulp with the same qualities as heavier paper—such as being able to print on both sides without the ink showing through, what printers call "opacity." When a harmless filler (calcium carbonate) is incorporated, or loaded, into the fibers, less wood fiber is needed to make the same amount of paper. Since the process is easier on fibers than mechanical pulping is, the resulting paper is also stronger and more recyclable. Because fiber is the most expensive material used in papermaking, and fiber-loading technology is quite inexpensive, mills that retrofit their old equipment could pay for their investment in one year on the material and energy savings alone. Fiber loading may be an especially good way to upgrade mixed office paper waste into high-quality paper, and reduce the weight of newspaper without sacrificing quality. For newspaper manufacturing, for instance, about 15 percent less wood fiber is needed (because the filler substitutes for wood and the weight of the paper can be reduced).[106]

A major focus of effort by regulators, industry, and environmentalists in most industrial nations continues to be on cleaning up the bleaching process. Reducing or totally eliminating chlorine from processing has a number of advantages, both ecological and economic. Chlorine is corrosive to processing equipment (which limits its recyclability within a plant and requires more expensive equipment), dangerous to workers, and extremely harmful when released to the environment. But there are ways to eliminate its use.

"Totally chlorine-free" (TCF) processes replace chlorine-based chemicals with oxygen-based chemicals such as hydrogen peroxide. With TCF systems, dioxin releases are virtually eliminated, almost no hazardous air pollutants are released, less water is needed, and that water can be reused many times (an essential step toward developing closed-loop systems). The water that is discharged needs less treatment, and energy use is far lower. All of these advances increase profitability. Chlorine-free processes are also safer for workers than processes using chlorine compounds, which are highly unstable, making them explosive and deadly. Scandinavia has already shifted 27 percent of its production to TCF. The United States and Canada lag far behind with less than 1 percent. TCF mills require less capital to build because they need less equipment and can use less expensive metals than mills using chlorine—an important benefit when new capacity is added. Malaysia has opened a new chlorine-free, recycled newsprint mill that will meet 75 percent of that nation's newsprint need. Worldwide, about 6 percent of the world's bleached pulp is totally chlorine free.[107]

So far, most of the industry (especially in North America) has chosen to adopt the "elemental chlorine-free" (ECF) approach rather than TCF because it requires fewer modifications to existing plants. ECF uses chlorine derivatives (such as chlorine dioxide) rather than elemental chlorine, which reduces some—but not all—of the most harmful compounds in the effluent. Worldwide, 54 percent of the bleached pulp produced in 1998 was ECF, a big jump from 17 percent the year before.[108]

Mills that retrofit could pay for their investment in one year.

Adoption of these newer bleaching methods has resulted in significant reductions in toxic discharges. An average paper mill using standard chlorine bleaching releases about 35 tons of organochlorines a day, while ECF mills release seven to 10 tons. TCF mills, on the other hand, produce and release none. In Scandinavia, where most of the world's TCF mills are located, detectable chlorine and dioxin discharges have been eliminated. According to the North American pulp and paper industry, the adoption of ECF has reduced measurable dioxin discharges from pulp and paper mills there by 96 percent between 1988 and 1994. Still, *total* effluent flows from the best ECF mills are twice as high as those from TCF mills.[109]

A potential opportunity for speeding up the shift to the more advanced technology in North America was missed when the U.S. EPA recently named ECF, rather than TCF, the "best available technology" under new combined air/water regulations for pulp and paper mills (the Cluster Rule). EPA also decided not to include oxygen delignification, another way to reduce the amount of bleaching and energy needed. Under pressure from industry, the EPA rejected the better technology, saying that TCF bleaching fails to meet the brightness level required for the market, a claim belied by the many TCF mills that exceed the standard. And in Canada, a 1992 British Columbia law gave pulp and paper mills 10 years to achieve the goal of completely eliminating organochlorines from their discharge. Rather than do so by switching to TCF processes, the industry has largely opted to go to ECF and is demanding that the law be overturned.[110]

The reluctance to move to the full range of cleaner technologies is surprising given their proven cost benefits. A 1995 study of 50 mills in six countries found that the earlier a mill had invested in improved technologies (such as extended delignification, ECF, and TCF), the more profitable and competitive it became. These findings held true even in nations without strong pollution regulations. As the study published in the industry journal *Pulp and Paper International* noted, "Business people have been brainwashed by classes in traditional economics to believe that investing to reduce pollution is a waste of money. The problem with this view is that it makes assumptions that do not hold true in the real world." The traditional view assumes that preventing pollution neither saves money nor increases output. But the reality is that such investments lower operating costs. As new capacity is added around the world, mills that shun cleaner technologies will find themselves at a competitive disadvantage.[111]

Designing a Sustainable Paper Economy

As the world enters a new century, the need for a healthier paper diet is more urgent than ever. In its most recent projections, the FAO predicts that global paper consumption will reach nearly 391 million tons by 2010, up more than 30 percent over today's level. With such growth would come the felling of more trees, and more pollution and waste. But ever-rising consumption and all the costs it entails are not a foregone conclusion. Accelerating the use of promising technologies and promoting leaner consumption habits can bring a healthier paper diet within reach.[112]

Paper use has traditionally been closely correlated with income levels, and industry analysts generally treat rising consumption as a sign of a healthy economy and improved quality of life. But higher paper consumption is no more a prerequisite for rising economies and standards of living than are increased automobile traffic, television viewing, pollution, or heart disease—other trends that historically have grown with income. Paper use can be decoupled from economic growth, just as energy use has been in most industrial countries thanks to strides in technological efficiency and conservation.[113]

In many parts of the world, expanded access to paper is needed if education, communication, and sanitation are to improve. About 80 percent of the world's people consume under 40 kilograms per person per year. But basic needs for paper will have to be met in an environmentally sound way lest the burden of forest degradation, air and water pollution, and waste generation outweigh the benefits they provide. This means embracing the best forest and industrial practices and rejecting the old model of ever-rising consumption.[114]

The high consumption levels prevalent in most industrial nations can be substantially reduced without losing the benefits of the services that paper now provides. Ways to reduce and reuse material have been demonstrated. About 85 to 90 percent of paper use is ephemeral, and nearly half of all paper is used for packaging. Recycled material already provides 38 percent of the fiber supply for paper, but this amount could be increased substantially. And even before being recycled, many paper products can be reused far more than they are today.[115]

Businesses can become more "eco-efficient" by adding value to products and services while reducing material and energy use, pollution, and waste. Today's box manufacturers, for instance, could sell shipping services, as some are already doing by selling or leasing reusable containers. Companies could sell information transmittal or communications services rather than reams of paper and copy machines. Businesses that move away from the strategy of "more material throughput equals more profit" will find that they can become more profitable by using and wasting less material while providing better services.[116]

Each of these strategies can provide considerable benefits. When combined, they can yield substantial savings by "capturing the magic of compounding arithmetic," as Hawken, Lovins, and Lovins describe in their book *Natural Capitalism*. If, for example, a process has 10 steps, and you can save 20 percent at each step, the net savings is a whopping 89 percent. The highest gains are made by reducing consumption because the impact of that savings is multiplied back through the entire chain of production, from the mills to the forests.[117]

Applying the principle of compounding arithmetic to paper consumption and production, dramatic savings could be achieved using some fairly conservative estimates of improvements along the chain. For example, if per capita consumption in today's high-consuming industrial countries were trimmed by one third—an amount possible largely through "good housekeeping" alone—global consumption would fall by 5 percent. At the same time, developing-country consumption could rise to 30 kilograms per person per year, enhancing the ability to meet basic needs.[118]

If industrial countries trimmed consumption by one third *and* production efficiency increased by 5 percent *and* recycled paper as a fiber source expanded to 60 percent (from today's 38 percent) *and* nonwoods as a fiber source doubled, total global consumption would fall, developing-country consumption could rise, *and* 56 percent of the wood fiber now used for papermaking could be saved. Beyond the obvious benefit of reducing paper's impact on the world's forests would also come substantial reductions in the other burdens of paper production and consumption—pollution, energy and water use, waste disposal, and the monetary cost of paper use.[119]

The action of policymakers is essential to making the transition to a sustainable paper economy. Through policies and regulations, governments can spur cleaner production, encourage recycling, promote sound forest management, shift consumption patterns, and save taxpayers money.

Fiscal reform is a top priority, starting with reducing the extensive subsidies for use of raw materials (such as wood, water, and energy), tax incentives for forest conversion, and low concession and stumpage fees. Many countries rely on such policies in the mistaken assumption that forest exploitation, trade, and increased consumption are engines for economic growth. In addition to being wasteful in their own right, these policies put relatively sustainable alternatives and activities like recycling and nonwood fiber sources at an economic disadvantage. The overall effect is to make paper produced from virgin wood fiber artificially inexpensive, encouraging overproduction and wasteful consumption.

Eliminating numerous widespread subsidies would yield substantial financial and environmental benefits. In Canada—the world's largest timber exporter—the province of British Columbia subsidizes the timber industry to the tune of $7 billion a year. Indonesia's subsidies for forest exploitation range between $1 and $3 billion a year. Some subsidies are direct, such as the $811 million in tax breaks the U.S. government annually gives the forest industry. Others are indirect, such as tax-exempt bonds for landfills and incinerators, and the subsidized energy, water, and transportation infrastructure that make virgin extraction and processing more profitable. The pulp and paper industry is the biggest beneficiary of one such project, the U.S. taxpayer-funded $2 billion Tennessee-Tombigee waterway, which allows access to previously landlocked forests and has made the explosive growth in wood chip exports possible.[120]

Taxes can be an effective tool for shifting the industry in a more sustainable direction. Higher taxes for exceeding pollution levels and for excess packaging, waste incineration, and landfilling have the double benefit of discouraging things people want less of (such as pollution) while lowering other taxes and thus encouraging things people want more of (such as jobs and investment). By gradually increasing taxes on water pollution in recent decades, the Netherlands has prodded industry to adopt cleaner technologies which in turn have led to major improvements in water quality. Producers who neglected to innovate and just passed the cost of the taxes on to consumers found that customers switched to less expensive and less-polluting products.[121]

Most governments have been pursuing a policy of trade promotion and liberalization that encourages forest degradation. Current global trade agreements under negotiation are set to further reduce tariffs on forest products (potentially stimulating production) and possibly to eliminate non-tariff measures such as raw log export bans, eco-labelling, and health and safety rules. Some national laws covering packaging, recycling, certification, and eco-labelling are targeted for elimination under expanding global free trade rules because the World Trade Organization (WTO) views them as non-tariff barriers to trade. Also under attack are national sanitary standards intended to prevent the accidental introduction of exotic pests and diseases that can hide in wood products and decimate the forests of the importing nation. Governments should resist allowing international trade agreements to undermine progress that has been made in allowing consumers choice, promoting sound forest management, and stemming the tide of invasive species.[122]

Paradoxically, allowing trade policies to be interpreted in ways that constrain labeling efforts, and thus limit consumer access to information about products (such as how they were produced, whether or not they were harvested destructively or by illegal means or even using forced labor) is antithetical to free trade's own ideology that markets should be ruled by consumer choice. If such rules had been in effect in the 1980s, the international consumer boycotts that were so effective in helping topple South Africa's apartheid rule would have been declared illegal.

In addition to reforming unsustainable policies, scaling up successful policies with proven benefits could help improve the paper diet. Laws mandating waste reduction and recycling have expanded the supply of used paper available for paper manufacturing. German laws aimed at reducing excess packaging and increasing the amount recycled have succeeded in boosting paper recovery rates. In many countries, "pay-as-you-throw" programs provide an incentive to generate less trash and to recycle more by charging households only for the amount they throw away, much the way many businesses already pay trash haulers based on volume. Policies that encourage recycling mills to locate in urban areas close to the fiber sources and markets, or agricultural fiber mills to set up in areas where such fibers are underutilized, can simultaneously make use of wasted resources and provide regional economic benefits. Ordinances that encourage the procurement of recycled or totally chlorine-free paper have stimulated the markets for more environmentally friendly paper.[123]

Even taking into account some successes in reducing pulp and paper pollution in many countries, room for improvement exists. Regulators, as well as industry, can use more integrated approaches to pollution prevention and control, source reduction, and the more prudent use of raw materials. Such an initiative is set to come into force in Europe. Many countries lack strong pollution controls or standards for industrial energy and water use and could benefit from them.[124]

Sometimes sound policies and regulations fail to translate into action because of poor implementation or enforcement. The U.S. government, whose purchases account for 2 percent of the U.S. paper market, could further stimulate progress by fully implementing the 1993 and 1998 executive orders mandating levels of post-consumer recycled content in government paper purchases. While a few agencies are leading the way, many others have very low compliance. Regulators may also bow to pressure by industry, as is happening to British Columbia's pollution policy mandating zero discharge of deadly dioxins. The efforts of many developing countries to enforce pollution control measures or halt illegal logging and land conversion are often hampered by understaffed and underfunded environmental agencies. Good laws may also be stymied by corruption at various levels of government.[125]

Relatively small investments in research by government and industry can yield big dividends. Given the economic importance of paper, surprisingly little is spent by government and industry on research and development. The pulp and paper industry spends less on research than any other major industry, about 1 percent of sales (compared to 4 to 5 percent in manufacturing industries as a whole). Even the USDA Forest Products Lab, which has sparked many innovations, has undergone repeated budget cuts. A hopeful sign is that several developing nations, notably India, Indonesia, and China, have established research institutes that focus on the environmental problems of their pulp and paper industries.[126]

By funding innovative research efforts and the adoption of sustainable production technologies and forest management, international lenders can help industries and nations make much-needed shifts to economic and environmental sustainability. Often, however, their financing has the opposite effect. The International Monetary Fund's policies after the Asian economic crisis, for example, encouraged Indonesia to expand its pulp and palm oil plantations, despite the well-known fact that these activities were largely responsible for Indonesia's devastating forest fires in 1997 and 1998.[127]

Because of its phenomenal utility, paper has become a fundamental and ubiquitous part of the global economy. In many paper-rich countries, people have grown so accustomed to the availability of cheap paper that it is hard to imagine sharing a thought with colleagues without printing out a memo on the whitest and brightest paper available, or sending a child off to school without single-serving snacks encased in little cardboard boxes. But the costs of ever-expanding production and consumption of paper are becoming increasingly untenable. The sooner we shift these trends, the greater the benefits will be—in terms of trees and soils saved, water and air pollution avoided, landfill and incinerator pressures eased, and money saved. It is fortunate that the ability to design a less damaging paper economy is at hand, for managing our relationship with this apparently "ordinary" material is essential to achieving an environmentally sustainable society in the new century.

Notes

1. Japanese banks from Public Radio International, "Marketplace Morning Report," 20 September 1999; special editions and supplements from Mikael Selling, "Producers Look to Benefit from the Millennium Effect," *Pulp and Paper International,* online edition, 1 August 1999.

2. Per capita levels from Miller Freeman, Inc., *International Fact and Price Book 1999* (San Francisco: 1998).

3. U.K. consumption of 13 kilograms per capita from Nick Robins and Sarah Roberts, "Rethinking Paper Consumption," International Institute for Environment and Development (IIED), September 1996, <www.iied.org/scati/pub/rethink.html>, viewed 28 September 1999.

4. Sixfold increase calculated from 46 million tons in 1950 from IIED, *Towards a Sustainable Paper Cycle* (London: 1996) and 299 million tons in 1997 from Miller Freeman, Inc., op. cit. note 2; projected demand in 2010 from Shushuai Zhu, David Tomberlin, and Joseph Buongiorno, *Global Forest Products Consumption, Production, Trade and Prices: Global Forest Products Model Projections to 2010* (Rome: Forest Policy and Planning Division, United Nations (U.N.) Food and Agriculture Organization (FAO), December 1998).

5. Per capita figures from Miller Freeman, Inc., op. cit. note 2; 30 to 40 kilograms from Mark P. Radka, "Policy and Institutional Aspects of the Sustainable Paper Cycle—An Asian Perspective," (Bangkok: U.N. Environmental Program (UNEP), Regional Office for Asia and the Pacific, 1994).

6. Wood Resources International, Ltd., *Fiber Sourcing Analysis for the Global Pulp and Paper Industry* (London: IIED, September 1996), reported that in 1993 wood pulp production required approximately 618 million cubic meters of wood—equal to 18.9 percent of the world's total wood harvest in 1993 (wood harvest volume as noted by FAO, *FAOSTAT Statistics Database,* <apps.fao.org>, viewed 30 October 1999); high ranking in resource use and pollution from IIED, op. cit. note 4, and Paper Task Force (Duke University, Environmental Defense Fund (EDF), Johnson & Johnson, McDonald's, Prudential Insurance Company of America, Time Inc.), *Paper Task Force Recommendations for Purchasing and Using Environmentally Preferable Paper* (New York: EDF, 1995); Franklin Associates, Ltd., "Characterization of Municipal Solid Waste in the United States: 1998 Update," report prepared for the U.S. Environmental Protection Agency (EPA), Municipal and Industrial Solid Waste Division, Office of Solid Waste, Washington, DC, July 1999.

7. Forty-three percent recycled from Miller Freeman, Inc., op. cit. note 2; energy and water use reduction from United States-Asia Environmental Partnership (USAEP) and the Civil Engineering Research Foundation (CERF),

"Clean Technologies in U.S. Industries: Focus on Pulp and Paper" (Washington, DC: September 1997).

8. Per capita from Miller Freeman, Inc., op. cit. note 2.

9. Savings of one third a Worldwatch calculation using FAO projected demand in 2010 from Zhu et al., op. cit. note 4.

10. Advertisement for UPM-Kymmene, *Financial Times*, 7 December 1988.

11. Four hundred fifty grades from IIED, op. cit. note 4.

12. Two hundred ninety-nine from Miller Freeman, Inc., op. cit. note 2; 1950 from IIED, op. cit. note 4; projected demand in 2010 from Zhu et al., op. cit. note 4. "Industrial" here means "developed" in the FAO categories.

13. Tables 1 and 2 from Miller Freeman, Inc., op. cit. note 2; growth in developing countries and Figure 1 from FAO, op. cit. note 6, viewed 28 October 1999; projections to 2010 from Zhu et al., op. cit. note 4. "Industrial" here means "developed" in the FAO categories.

14. Per capita figures and Figure 2 from Miller Freeman, Inc., op. cit. note 2; developing- and industrial-country figures from FAO, op. cit. note 6, viewed 21 October 1999.

15. Thirty- to 40- kilogram estimate from Radka, op. cit. note 5.

16. Miller Freeman, Inc., op. cit. note 2; Figure 3 from FAO, op. cit. note 6.

17. Replacement of paper with plastic from Miller Freeman, Inc., *1999 North American Pulp and Paper Factbook* (San Francisco: 1998); tripling from FAO, op. cit. note 6, viewed 22 October 1999.

18. Paper.com, "Surprising Many, Paper Use Soars with Internet Growth," press release, <www.papercom.org/press6.htm>, viewed 22 April 1999.

19. Paper use in offices from Miller Freeman, Inc., op. cit. note 17, and Raju Narisetti, "Pounded by Printers, Xerox Copiers Go Digital," *Wall Street Journal*, 12 May 1998; U.S. Civilian Labor Force from Bureau of Labor Statistics, *Employment and Earnings*, January 1999, <www.bls.gov/cpsaatab.htm>, viewed 15 September 1999; U.S. mail from Direct Marketing Association, *Statistical Factbook 1999* (New York: 1999); number of households from U.S. Census Bureau, "Household and Family Characteristics," P20-515 (Suitland, MD: 1998).

20. See commentary in Miller Freeman, Inc., op. cit. note 17, and Lauri Hetemäki, "Information Technology and Paper Demand Scenarios," in Matti Palo and Jussi Uusivuori, eds., *World Forests, Society and Environment*, (Dordrecht, Netherlands: Kluwer Academic Publishers, 1999).

21. Figure 4 from FAO, op. cit. note 6.

22. Ten percent from Gary Mead, "Tough Year Ahead for Pulp and Paper," *Financial Times*, 31 January 1998; growth in various countries from Miller Freeman, Inc., op. cit. note 2.

23. Hou-Min Chang, "Economic Outlook for Asia's Pulp and Paper Industry," *TAPPI Journal*, January 1999.

24. Dorothy A. Paun et al., "A Performance Analysis of North American Pulp, Paper, and Packaging Companies," *TAPPI Journal*, December 1998; Harold M. Cody, "Latin American Demand, Trade and Global Competition Expanding," *Pulp and Paper Magazine*, 1 October 1998; Greg McIvor, "North Moves South and West Moves East," *Financial Times*, 7 December 1998.

25. Amount traded from FAO, op. cit. note 6; 45 percent from Bruce Michie, Cherukat Chandrasekharan, and Philip Wardle, "Production and Trade in Forest Goods," in Palo and Uusivuori, op. cit. note 20.

26. Ten and 46 percent from the American Forest and Paper Association (AF&PA), confirmed by Michael Klein, personal communication with Ashley Mattoon, 20 October 1999; 24 percent from Paun et al., op. cit. note 24; Patrick Knight, "Brazil Cautiously Optimistic amid the Chaos," *Pulp and Paper International*, April 1999; export destinations from Miller Freeman, Inc., op. cit. note 2; pulp exports from Mark Payne, "Latin America Aims High for the Next Century," *Pulp and Paper International*, August 1999.

27. Growth calculated from data in Miller Freeman, Inc., op. cit. note 2; projections for 2010 from Xiang-Ju Zhong, "Challenges and Opportunities in China," *Pulp and Paper International*, August 1998.

28. Zhang Yan, "Reliance on Foreign Lumber to Increase," *China Daily*, 7 September 1998; 40 million from Bruce Gilley, "Sticker Shock," *Far Eastern Economic Review*, 14 January 1999; imports from FAO, op. cit. note 6.

29. McIvor, op. cit. note 24; Phil Crawford, "Global Pulp and Paper Industry in Transition," *TAPPI Journal*, January 1999.

30. One mill can affect the entire market from Crawford, ibid.

31. Ten percent from Paul Hawken, Amory Lovins, and Hunter Lovins, *Natural Capitalism: Creating the Next Industrial Revolution* (New York: Little Brown, 1999).

32. Average annual natural forest cover loss between 1990 and 1995 was 13.7 million hectares according to FAO, *State of the World's Forests 1999* (Rome: 1999); causes of degradation from Dirk Bryant, Daniel Nielsen, and Laura Tangley, *The Last Frontier Forests, Ecosystems and Economies on the Edge* (Washington, DC: World Resources Institute (WRI), 1997).

33. Amount of wood harvest going to paper from Wood Resources International, op. cit. note 6; 41.8 percent calculated by using the world's total industrial wood harvest as noted by FAO, op. cit. note 6, viewed 30 October 1999; paper growing twice as fast as other major wood products from FAO, op. cit. note 32; 2050 from World Commission on Forests and Sustainable Development, *Our Forests Our Future* (Cambridge, U.K.: Cambridge University Press, 1999).

34. Portion from mill residues from Wood Resources International, op. cit. note 6, which reported that 63 percent of the 618 million cubic meters was from the roundwood pulpwood supply and 37 percent was from manufacturing residues and off-site chipping operations.

35. Proportions of wood, recycled, and nonwood sources and Figure 6 from FAO, op. cit. note 6, viewed 29 October 1999; wood fiber sources in Figure 5 from IIED, op. cit. note 4.

36. Growth rates from Rita Pappens, "Chile Faces Up to Financial and Forest Challenges," *Pulp and Paper International*, April 1999; cost from Nelson Noel, "Paper and Forest Products Industry Outlook" (New York: Moody's Investors Service, April 1999).

37. IIED, op. cit. note 4.

38. Trends in fiber supply from plantations from FAO, op. cit. note 32; Sustainable Forestry Working Group, ed., *The Business of Sustainable Forestry: Case Studies* (Chicago: John D. and Catherine T. MacArthur Foundation, 1998); subsidized plantation and foreign investment from Ricardo Carrere and Larry Lohmann, *Pulping the South: Industrial Tree Plantations and the World Paper Economy* (London: Zed Books, 1996); Japan from Japan Paper Association (JPA), *Pulp and Paper Statistics 1998* (Tokyo: 1999); JPA, *In Harmony with Nature* (Tokyo: October 1997).

39. Estimate of 13 million calculated from 10 percent of current plantation estate from World Commission on Forests and Sustainable Development, op. cit. note 33; current plantation estate of approximately 130 million hectares (calculated by adding the estimate of 60 million hectares in industrial countries to the 70 million in developing countries) from FAO, op. cit. note 32; Brazil and Chile from Jeremy Williams, *A Study of Plantation Timber Prices in Latin America and the Southern United States of America* (Rome: FAO, 1998).

40. Benefits of plantations from FAO, op. cit. note 32; Roger Sedjo and Daniel Botkin, "Using Forest Plantations to Spare Natural Forests," *Environment*, December 1997; World Commission on Forests and Sustainable Development, op. cit. note 33.

41. 1.4 million from IIED, op. cit. note 4; 80 percent from Nigel Dudley, *The Year the World Caught Fire* (Gland, Switzerland: World Wide Fund for

Nature, December 1997); Fred Pearce, "Playing with Fire," *New Scientist*, 21 March 1998.

42. M.J. Mac et al., *Status and Trends of the Nation's Biological Resources*, vol. 1 (Reston, VA: U.S. Department of the Interior, U.S. Geological Survey, 1998); Sten Nilsson et al., "How Sustainable Are North American Wood Supplies?" Interim Report (Laxenburg, Austria: International Institute for Applied Systems Analysis, 1999); U.S. Forest Service data from U.S. Department of Agriculture (USDA), Forest Service, *The South's Fourth Forest: Alternatives for the Future*, Forest Resource Report no. 24 (Washington, DC: June 1988).

43. Indonesia from Lesley Potter and Justin Lee, *Tree Planting in Indonesia: Trends, Impacts and Directions*, Occasional Paper no. 18 (Bogor, Indonesia: Center for International Forestry Research, December 1998); Tupinikim and Guarani from Carrere and Lohmann, op. cit. note 38; World Rainforest Movement (WRM), *Tree Plantations: Impacts and Struggles* (Montevideo: WRM, February 1999).

44. Shortage of arable land from Gary Gardner, *Shrinking Fields: Cropland Loss in a World of Eight Billion,* Worldwatch Paper 131 (Washington, DC: Worldwatch Institute, July 1996); 90 million from FAO, *State of the World's Forests 1997* (Oxford, U.K.: 1997); 100 million hectares from World Commission on Forests and Sustainable Development, op. cit. note 33; FAO, op. cit. note 32.

45. Paper Task Force, op. cit. note 6; IIED, op. cit. note 4.

46. World Energy Council 1995 cited in IIED, op. cit. note 4; Paper Task Force, op. cit. note 6.

47. Japan from JPA, op. cit. note 38, and IIED, op. cit. note 4; U.S. from Miller Freeman, Inc., op. cit. note 17.

48. Liters for virgin fiber paper from Paper Task Force, op. cit. note 6; USAEP and CERF, op. cit. note 7.

49. USAEP and CERF, op. cit. note 7; JPA, op. cit. note 38.

50. Miller Freeman, Inc., op. cit. note 17.

51. Pulp process from Miller Freeman, Inc., op. cit. note 17, and IIED, op. cit. note 4; pulp rations from FAO, op. cit. note 6.

52. U.S. from IIED, op. cit. note 4; Asia from Radka, op. cit. note 5.

53. Miller Freeman, Inc., op. cit. note 17; Reach for Unbleached!, "Sludge Spreading a Problem in Ontario and BC Too," press release (Whaletown, BC: 7 May 1999), <www.rfu.org>, viewed 12 November 1999.

54. Miller Freeman, Inc., op. cit. note 17.

55. IIED, op. cit. note 4.

56. IIED, op. cit. note 4; Miller Freeman, Inc., op. cit. note 17.

57. IIED, op. cit. note 4; Miller Freeman, Inc., op. cit. note 17; U.S. EPA, "Fact Sheet: EPA's Final Pulp, Paper, and Paperboard 'Cluster Rule'—Overview," November 1997, EPA-821-F-97-010; Bill Nichols, "Four Years of Work, Debates Produce First Phase of EPA's Cluster Rule," *Pulp and Paper*, January 1998.

58. Radka, op. cit. note 5.

59. Twenty percent from Nick Robins and Sarah Roberts, "Rethinking Paper Consumption," IIED, September 1996 <www.iied.org/scati/pub/rethink.html>, viewed 28 September 1999; U.S. EPA, WasteWise Program Updates (various) <www.epa.gov/wastewise>; Bruce Nordman, U.S. Department of Energy, Lawrence Berkeley National Laboratory, <eetd.lbl.gov.paper>, viewed 2 August 1999; Xerox, "Xerox Business Guide to Waste Reduction and Recycling," <www.xerox.com>, viewed 4 November 1999.

60. Nordman, op. cit. note 59. (American 20-pound paper is 75.2 g/m2; Europe uses 80 g/m2, Japan uses 65 g/m2).

61. Bank of America, *1997 Environment Report*, <www.bankamerica.com/environment>, viewed 17 August 1999.

62. Weight reductions from Nordman, op. cit. note 59; size reductions from Susan Kinsella, "Recycled Paper Buyers, Where Are You?" *Resource Recycling*, November 1998.

63. Jerry Powell, "Seven Hot Trends in Paper Recycling," *Resource Recycling* (Recovered Paper Supplement), April 1998.

64. United Parcel Service (UPS) and Alliance for Environmental Innovation (a project of the EDF and Pew Charitable Trusts), "Achieving Preferred Packaging: Report of the Express Packaging Project," (New York: November 1998); Elizabeth Sturcken, "Preferred Packaging: Accelerating Environmental Leadership in the Overnight Shipping Industry," report, Alliance for Environmental Innovation (New York: December 1997); "Federal Express Joins with Recycled Paperboard Alliance," *Pulp and Paper Online*, 25 October 1999, <www.pponline.com>.

65. Paper Recycling Promotion Center, "Manual for Used Office Paper Recycling [Case Studies and Success Stories]" (Tokyo: March 1999).

66. Paper Task Force, op. cit. note 6; Susan Kinsella, "Environmentally

Sound Paper Overview: The Essential Issues," *Conservatree's Greenline*, October 1996; Dan Imhoff, *The Simple Life Guide to Tree-Free, Recycled and Certified Papers* (Philo, CA: SimpleLife, 1999); Coop America, *Woodwise Consumer Guide* (Washington, DC: 1999), <www.woodwise.org>.

67. Janet Abramovitz, personal communication with Peter Ince, USDA, Forest Products Laboratory, 30 July 1999.

68. Procter and Gamble from IIED, op. cit. note 4.

69. U.S. EPA, *WasteWise Fifth Year Progress Report* and "Wastewise Update" June 1996, <www.epa.gov/wastewise>; Xerox, op. cit. note 59; Hawken, op. cit. note 31.

70. Nordman, op. cit. note 59; Nevin Cohen, "Greening the Internet: Ten Ways E-Commerce Could Affect the Environment," *Environmental Quality Management*, autumn 1999.

71. UPS from UPS and Alliance for Environmental Innovation, op. cit. note 64; U.S. EPA, op. cit. note 69.

72. Miller Freeman, Inc., op. cit. note 17.

73. World Wide Fund for Nature, Forests for Life Campaign, Buyers Groups Information and Contacts, <www.panda.org>, viewed 10 November 1999.

74. Paper Recycling Promotion Center, op. cit. note 65.

75. Paper Task Force, op. cit. note 45; UPS and Alliance for Environmental Innovation, op. cit. note 64; Sturcken, op. cit. note 64; U.S. EPA, op. cit. note 69; USDA FPL, *1999 Research Highlights* (Madison, WI: 1999).

76. Imhoff, op. cit. note 66; Paper Task Force, op. cit. note 46; Kinsella, op. cit. note 66; Coop America, op. cit. note 66; Chlorine Free Paper Association, "Guide to TCF & PCF Papers" (Algonquin, IL: 1997).

77. FAO, op. cit. note 6.

78. Ishiguro and Akiyama 1994 cited in IIED, op. cit. note 4; Allen Hershkowitz, *Too Good To Throw Away: Recycling's Proven Record* (New York: Natural Resources Defense Council, February 1997); Imhoff, op. cit. note 66.

79. Douglas J. Burke, "1st Urban Fiber—a Pulp Mill in the Urban Forest," *TAPPI Journal*, January 1997; Weld F. Royal, "Paper Mill Project Plants Roots in the South Bronx," *Biocycle*, July 1994; Herbert Mushcamp, "Greening a South Bronx Brownfield," *New York Times*, 23 January 1998; John Holusha, "A New Life for a Polluted Old Bronx Rail Yard," *New York Times*, 6 July 1997.

80. U.S. figure from Franklin Associates, Ltd., op. cit. note 6.

81. Rates from Miller Freeman, Inc., op. cit. note 2; volume from FAO, op. cit. note 6; projection from Zhu et al., op. cit. note 4.

82. Miller Freeman, Inc., op. cit. note 2; Paper Recycling Promotion Center, "Information about Paper Recycling Promotion Center" (Tokyo: n.d.).

83. Miller Freeman, Inc., op. cit. note 17; "Divided EU Agrees on Packaging Directive, Joint Ratification of Climate Change Treaty," *International Environment Reporter*, 12 January 1994; "Council Agrees on FCP Phase-out, Packaging Recycling, Hazardous Waste List," *International Environment Reporter*, 11 January 1995; EC Directive from "Report Highlights Benefits of Increasing Newspaper Recycling," *ENDS Report* 282, July 1998; Netherlands from "Paper, cardboard recycling proceeding well to meet 2001 targets, industry official says," *International Environment Reporter*, 2 September 1998.

84. Miller Freeman, Inc., op. cit. note 2; 1940s from Maureen Smith, *The U.S. Paper Industry and Sustainable Production* (Cambridge, MA: MIT Press, 1997); Table 3 from Miller Freeman, Inc., op. cit. note 2.

85. FAO, op. cit. note 6.

86. Fred D. Iannazzi, "A decade of progress in U.S. paper recovery," *Resource Recycling*, June 1999; Kinsella, op. cit. note 62; U.K. from "Report highlights," op. cit. note 83.

87. Price fluctuations, wastepaper supply, and pulping capacity from Miller Freeman, Inc., op. cit. note 17.

88. Lisa A. Skumatz and Robert Moylan, "How to Re-energize Recycling Progress," *Resource Recycling*, June 1999.

89. Some criticism of recycling includes John Tierney, "Recycling is Garbage," *New York Times Magazine*, 30 June 1996, and Fred Pearce, "Burn Me," *New Scientist*, 22 November 1997; rebuttals include Hershkowitz, op. cit. note 78; Lauren Blum, Richard A. Denison, and John F. Ruston, "A Lifecycle Approach to Purchasing and Using Environmentally Preferable Paper," *Journal of Industrial Ecology*, vol. 1, no. 3, 1997; Richard A. Denison and John F. Ruston, "Recycling Is Not Garbage," *Technology Review*, October 1997; and an extensive online forum on the *New Scientist*'s website, <www.newscientist.com>, viewed 28 October 1999.

90. Process improvements from Jim Kenny, "More Recycled Fibers Generate Process Improvements," *Pulp and Paper Magazine*, June 1999, and from Marguerite Sykes et al., "Environmentally Sound Alternatives for Upgrading Mixed Office Wastes," *Proceedings of the 1995 International Environmental Conference*, 7–10 May 1995, Atlanta, GA, TAPPI Press.

91. Smith, op. cit. note 84.

92. Tsuen-Hsuin Tsien, *Written on Bamboo and Silk, The Beginnings of Chinese Books and Inscriptions* (Chicago: University of Chicago Press, 1962).

93. Percentages from FAO, op. cit. note 6; closure of mills in China from FAO, op. cit. note 44; China's growing dependence on wood from "Asia 'Miracle' Will Continue in Fiber," *International Woodfiber Report*, November 1997; Xiang-Ju Zhong, op. cit. note 27.

94. James S. Han, "Properties of Nonwood Fibers," in *Proceedings of the Korean Society of Wood Science and Technology Annual Meeting*, 1998 (Seoul: Korean Society of Wood Science and Technology, 1998).

95. Silica content from ibid; Smith, op. cit. note 84; 50 percent from soil residues from IIED, op. cit. note 4; alternative techniques from FAO, op. cit. note 32.

96. Seventy-three percent of supply from Leena Paavilainen, "European Prospects for Using Nonwood Fibers," *Pulp and Paper International*, June 1998; straw yields from Smith, op. cit. note 84; total available from George A. White and Charles G. Cook, "Inventory of Agro-Mass," in Roger M. Rowell, Raymond A. Young, and Judith K. Rowell, eds., *Paper and Composites from Agro-Based Resources* (Boca Raton, FL: CRC Press, 1997); Joseph E. Atchison, "Update on Global Use of Non-wood Plant Fibers and some Prospects for Their Greater Use in the United States," in Technical Association of the Pulp and Paper Industry (TAPPI), *1998 TAPPI Proceedings North American Nonwood Fiber Symposium*, Atlanta, GA, 17–18 February 1998.

97. Burning of residues from Vaclav Smil, "Crop Residues: Agriculture's Largest Harvest," *Bioscience*, April 1999; David O. Hall et al., "Biomass for Energy: Supply Prospects," in Thomas B. Johansson et al., eds., *Renewable Energy: Sources for Fuels and Electricity* (Washington, DC: Island Press, 1993).

98. Nine percent from Paavilainen, op. cit. note 96; less energy and chemicals from Han, op. cit. note 94, and IIED, op. cit. note 4.

99. Sustainable Forestry Working Group, ed., op. cit. note 38; Brazil from Williams, op. cit. note 39.

100. "Company Watch," *Business Ethics*, May/June 1998; Pulp and Paper Online, "McDonald's calls for disclosure of forestry practices," <www.pponline.com/inside/stories/wk)03_08_1999/12htm>, viewed 9 March 1999.

101. World Wide Fund for Nature,"Certification Hits More Than 15 Million Hectares Worldwide," press release, 21 January 1999, <www.panda.org>; first FSC pulp from the Wilderness Society, "New York Mill Produces North America's First 'Green Certified' Paper," press release, Washington, DC, 15 October 1998.

102. Some examples of industry programs include the AF&PA's Sustainable Forestry Initiative and the Canadian Standards Association's Sustainable Forest Management Standard, AF&PA <www.afandpa.org/forestry/sfi_frame.html>, and Canadian Sustainable Forestry Certification Coalition, <www.sfms.com>; World Wildlife Fund, "Nation's Top Environmental Organizations Demand an End to Timber Industry Association's Greenwashing," press release, 29 April 1999, <www.worldwildlife.org>.

103. Blum et al., op. cit. note 89.

104. New Mexico mill in David J. Bentley, Jr., "McKinley Paper Goes From Zero to Success in Less than Four Years," *TAPPI Journal*, January 1999; "Green Bay Studying Plan to Build Series of Regional Mini-mills Based on Recycled Fiber," *Pulp and Paper Week*, 24 May 1993; "Case Closed," *Boxboard Containers*, October 1993.

105. IIED, op. cit. note 4; Paper Task Force, op. cit. note 6; David J. Senior et al., "Enzyme Use Can Lower Bleaching Costs, Aid ECF Conversions," *Pulp and Paper*, July 1999; 80 percent in Jay Ritchlin and Paul Johnston, "Zero Discharge: Technological Progress Towards Eliminating Kraft Pulp Mill Liquid Effluent, Minimizing Remaining Waste Streams and Advancing Worker Safety," prepared for Reach For Unbleached!, the Zero Toxics Alliance Pulp Caucus, and Greenpeace International, Reach for Unbleached! Foundation, Whaletown, BC, n.d.

106. USDA Forest Service, Forest Products Lab (FPL), "Biopulping: Technology Learned from Nature That Gives Back to Nature," (Madison, WI: March 1997); Masood Akhtar et al., "Toward Commercialization of Biopulping," *Paper Age*, February 1997; Gary Myers, USDA FPL, personal communication with Janet Abramovitz, 30 July 1999; fiber loading from John H. Klungness et al., "Lightweight, High Opacity Paper: Process Costs and Energy Use Reduction," presented at the AIChE Symposium/TAPPI Pulping Conference, 29 October 1998, Montreal; Oliver Heise et al., "Industrial Scale-up of Fiber Loading on Deinked Wastepaper," TAPPI Pulping Conference Proceedings, held 27–31 October 1996, Nashville, TN; Sykes et al., op. cit. note 90; John H. Klungness, research chemical engineer, USDA FPL, personal communication with Janet Abramovitz, 30 July 1999.

107. TCF advantages from Ritchlin and Johnston, op. cit. note 105, and Paper Task Force, op. cit. note 6; production from Alliance for Environmental Technology, "Trends in World Bleached Chemical Pulp Production: 1990–1998," <www.act.org>; Greenpeace, "Chlorine Use Sections—Pulp and Paper," <www.greenpeace.org/toxics/ci>, viewed 16 July 1999.

108. Ibid.; "MNI-Malaysia's International Class Newsprint Mill Fully Operational in April 1999," *Malaysia Timber Bulletin*, vol. 4, no. 10, 1998.

109. Release amounts from Greenpeace, "Alternatives to Chlorine TCF vs

ECF," <www.greenpeace.org/toxics/cp/cp-alts-pp3.html>, viewed 16 July 1999; 96 percent from Miller Freeman, Inc., op. cit. note 17; total effluent from Ritchlin and Johnston, op. cit. note 105.

110. "Elemental Chlorine-Free Bleaching Leads World Market in Pulp Production," *International Environment Reporter*, 11 November 1998; Todd Paglia, "The Clinton-Industry Cluster: EPA and Business Assure a Future of Dirty Paper Making," *Multinational Monitor*, January/February 1998; U.S. EPA, op. cit. note 57; "Cluster Rule Final; EPA Backs Industry Wish to Cut Air/Water Toxins Using ECF," *Pulp and Paper Week*, 17 November 1997; BC from Miller Freeman, Inc., op. cit. note 17; and "Zero AOX Law faces behind-the-scenes attack," Reach for Unbleached! press release, 7 May 1999, <www.rfu.org>.

111. Mill study in Chad Nehrt, "Spend More to Show Rivals a Clean Pair of Heels," *Pulp and Paper International*, 1 June 1995; IIED, op. cit. note 4; Paper Task Force, op. cit. note 6.

112. Projections from Zhu et al., op. cit. note 7.

113. Smith, op. cit. note 84; International Energy Agency, *Indicators of Energy Use and Efficiency: Understanding the Link Between Energy and Human Activity* (Paris: Organization for Economic Co-operation and Development (OECD), 1997).

114. Radka, op. cit. note 5.

115. Hawken et al., op. cit. note 31; 85/15 from Netherlands in "Paper, Cardboard Recycling Proceeding Well to Meet 2001 Targets, Industry Official Says," *International Environment Reporter*, 2 September 1998.

116. Stephen Schmidheiny, with Business Council for Sustainable Development, *Changing Course* (Boston: MIT Press, 1992); U.S. EPA, op. cit. note 69.

117. Hawken et al., op. cit. note 31.

118. Worldwatch Institute estimates based on FAO, op. cit. note 6; for 1997 actual consumption and 2010 projections (using FAO categories of developed and developing), see Zhu et al., op. cit. note 4.

119. Worldwatch estimates based on FAO, op. cit. note 6, and various estimates for potential improvements.

120. U.S. from Susan Kinsella, ed. "Welfare for Waste: How Federal Taxpayer Subsidies Waste Resources and Discourage Recycling," Grassroots Recycling Network, April 1999, <www.grrn.org>, and from Danna Smith, "Chipping Forests and Jobs: A Report on the Economic Impact of Chip Mills in the Southeast" (Brevard, NC: Dogwood Alliance and Native Forest Network,

August 1997); others from Nigel Sizer, David Downes, and David Kaimowitz, "Tree Trade: Liberalization of International Commerce in Forest Products: Risks and Opportunities" (Washington, DC: WRI and the Center for International Environmental Law, November 1999).

121. David Malin Roodman, *The Natural Wealth of Nations* (New York: W.W. Norton & Company, 1998); Duncan McLaren et al., *Tomorrow's World: Britain's Share in a Sustainable Future* (London: Earthscan, 1998).

122. Sizer et al., op. cit. note 120; Christopher Bright, "Invasive Species: Pathogens of Globalization," *Foreign Policy*, fall 1999.

123. Ruth Walker, "Fee-Based Recycling of Trash Is a Mixed Bag for Germans," *Christian Science Monitor*, 22 May 1997; U.S. EPA "Pay-As-You-Throw Introduction," <www.epa.gov/epaoswer/non-hw/payt>, viewed 13 April 1999; Smith, op. cit. note 84.

124. "IPPC Directive adopted, IPPC for small firms on the way," *ENDS Report* 261, October 1998; IIED, op. cit. note 4.

125. Kinsella, op. cit. note 62; Todd Paglia, "The Big Deal Recycling Executive Order," *Multinational Monitor*, January/February 1998; Radka, op. cit. note 5; BC from Reached for Unbleached!, op. cit. note 110.

126. Current and historic R&D spending from Miller Freeman, Inc., op. cit. note 17; comparison to R&D in other industries from Smith, op. cit. note 84; developing countries from Radka, op. cit. note 5.

127. International Monetary Fund (IMF), "Statement by the Managing Director on the IMF Program with Indonesia," Washington, DC, 15 January 1998; Sander Thoenes, "Indonesian Wood Cartel Resists IMF Reforms," *Financial Times*, 13 February 1998.

Worldwatch Papers

Worldwatch Papers by Janet N. Abramovitz and Ashley T. Mattoon

No. of Copies

____149. **Paper Cuts: Recovering the Paper Landscape** by Janet N. Abramovitz and Ashley T. Mattoon

____140. **Taking a Stand: Cultivating a New Relationship with the World's Forests** by Janet N. Abramovitz

____128. **Imperiled Waters, Impoverished Future: The Decline of Freshwater Ecosystems** by Janet N. Abramovitz

____148. **Nature's Cornucopia: Our Stake in Plant Diversity** by John Tuxill
____147. **Reinventing Cities for People and the Planet** by Molly O'Meara
____146. **Ending Violent Conflict** by Michael Renner
____145. **Safeguarding The Health of Oceans** by Anne Platt McGinn
____144. **Mind Over Matter: Recasting the Role of Materials in Our Lives** by Gary Gardner and Payal Sampat
____143. **Beyond Malthus: Sixteen Dimensions of the Population Problem** by Lester R. Brown, Gary Gardner, and Brian Halweil
____142. **Rocking the Boat: Conserving Fisheries and Protecting Jobs** by Anne Platt McGinn
____141. **Losing Strands in the Web of Life: Vertebrate Declines and the Conservation of Biological Diversity** by John Tuxill
____139. **Investing in the Future: Harnessing Private Capital Flows for Environmentally Sustainable Development** by Hilary F. French
____138. **Rising Sun, Gathering Winds: Policies to Stabilize the Climate and Strengthen Economies** by Christopher Flavin and Seth Dunn
____137. **Small Arms, Big Impact: The Next Challenge of Disarmament** by Michael Renner
____136. **The Agricultural Link: How Environmental Deterioration Could Disrupt Economic Progress** by Lester R. Brown
____135. **Recycling Organic Waste: From Urban Pollutant to Farm Resource** by Gary Gardner
____134. **Getting the Signals Right: Tax Reform to Protect the Environment and the Economy** by David Malin Roodman
____133. **Paying the Piper: Subsidies, Politics, and the Environment** by David Malin Roodman
____132. **Dividing the Waters: Food Security, Ecosystem Health, and the New Politics of Scarcity** by Sandra Postel
____131. **Shrinking Fields: Cropland Loss in a World of Eight Billion** by Gary Gardner
____130. **Climate of Hope: New Strategies for Stabilizing the World's Atmosphere** by Christopher Flavin and Odil Tunali
____129. **Infecting Ourselves: How Environmental and Social Disruptions Trigger Disease** by Anne E. Platt
____127. **Eco-Justice: Linking Human Rights and the Environment** by Aaron Sachs
____126. **Partnership for the Planet: An Environmental Agenda for the United Nations** by Hilary F. French
____125. **The Hour of Departure: Forces That Create Refugees and Migrants** by Hal Kane
____124. **A Building Revolution: How Ecology and Health Concerns Are Transforming Construction** by David Malin Roodman and Nicholas Lenssen
____123. **High Priorities: Conserving Mountain Ecosystems and Cultures** by Derek Denniston
____122. **Budgeting for Disarmament: The Costs of War and Peace** by Michael Renner
____121. **The Next Efficiency Revolution: Creating a Sustainable Materials Economy** by John E. Young and Aaron Sachs
____120. **Net Loss: Fish, Jobs, and the Marine Environment** by Peter Weber

____**Total copies (transfer number to order form on next page)**

PUBLICATION ORDER FORM

NOTE: Many Worldwatch publications can be downloaded as PDF files from our website at **www.worldwatch.org**. Orders for printed publications can also be placed on the web.

_____ *State of the World:* $13.95
The annual book used by journalists, activists, scholars, and policymakers worldwide to get a clear picture of the environmental problems we face.

_____ **State of the World Library: $30.00 (international subscribers $45)**
Receive *State of the World* and all five Worldwatch Papers as they are released during the calendar year.

_____ *Vital Signs:* $13.00
The book of trends that are shaping our future in easy-to-read graph and table format, with a brief commentary on each trend.

_____ WORLD WATCH **magazine subscription: $20.00 (international airmail $35.00)**
Stay abreast of global environmental trends and issues with our award-winning, eminently readable bimonthly magazine.

_____ **Worldwatch Database Disk Subscription: $89.00**
Contains global agricultural, energy, economic, environmental, social, and military indicators from all current Worldwatch publications. Includes a mid-year update, and *Vital Signs* and *State of the World* as they are published. Disk contains Microsoft Excel spreadsheets 5.0/95 (*.xls) for Windows.
Check one: _____ PC _____ Mac

_____ **Worldwatch Papers—See list on previous page**
Single copy: $5.00
2–5: $4.00 ea. • 6–20: $3.00 ea. • 21 or more: $2.00 ea.

$4.00* Shipping and Handling *($8.00 outside North America)*
minimum charge for S&H; call (800) 555-2028 for bulk order S&H

_____ **TOTAL** (U.S. dollars only)

Make check payable to: Worldwatch Institute, 1776 Massachusetts Ave., NW, Washington, DC 20036-1904 USA

Enclosed is my check or purchase order for U.S. $_____

☐ AMEX ☐ VISA ☐ MasterCard _____
                                    Card Number   Expiration Date

signature

name   **daytime phone #**

address

city   **state**   **zip/country**

phone: (800) 555-2028   fax: (202) 296-7365   e-mail: wwpub@worldwatch.org
                    website: www.worldwatch.org

Wish to make a tax-deductible contribution? Contact Worldwatch to find out how your donation can help advance our work.